D0209535

The Great Apes

The Great Apes

A Short History

Chris Herzfeld

Translated by Kevin Frey

FOREWORD BY JANE GOODALL

Yale
UNIVERSITY PRESS
New Haven & London

Yale University Press books may be purchased in quantity for educational, business, or promotional use. For information, please e-mail sales.press@yale.edu (U.S. office) or sales@yaleup.co.uk (U.K. office).

Set in Bulmer type by Newgen North America, Austin, Texas.
Printed in the United States of America.

ISBN 978-0-300-22137-4 (hardcover : alk. paper)
Library of Congress Control Number: 2017940993
A catalogue record for this book is available from the British Library.

This paper meets the requirements of ANSI/NISO Z39.48–1992 (Permanence of Paper).

10 9 8 7 6 5 4 3 2 1

Frontispiece: Victoria (June 9, 1968–May 21, 2016), Antwerp Zoo, 1996.
(This portrait appears in the book *Les grands singes: L'humanité au fond des yeux*, by Pascal Picq, Dominique Lestel, Vinciane Despret, and Chris Herzfeld [Editions Odile Jacob, 2005], © Editions Odile Jacob/Chris Herzfeld, MNHN, 2005)

Contents

Foreword

WE HAVE ALWAYS BEEN FASCINATED BY the great apes—orangutans, gorillas, bonobos, and chimpanzees. Hardly surprising since we are, in fact, the fifth great ape! Today we know quite a lot about these amazing beings—but when I first went to Tanzania to try to observe chimpanzees in the wild, almost nothing was known about any of the apes in their natural habitat with the exception of the mountain gorilla, which had been studied by George Schaller. And not much was known about ape behavior in captivity, either. I read all that I could find on the subject, and I went to look at the two pathetic individuals in the London Zoo. I learned most from a book called *The Mentality of Apes,* by Wolfgang Kohler, who worked with a colony of chimpanzees that had been established at Tenerife in the Canary Islands. His observations were extremely revealing.

It was in 1960 that I arrived in the Gombe National Park, site of what is now a field study of chimpanzees spanning nearly sixty years. Back then my biggest problem was that the shy apes ran off as soon as they saw me. But eventually, after several months, they began to lose their fear, and I was able to learn more and more about their fascinating social structure and way of life.

When I went to Gombe I had not been to college. But after I had been in the field for over a year my mentor, Louis Leakey, got me a place at Cambridge University to work for a PhD in ethology—he said there was no time to get a BA first, but that I would need a PhD if I was ever to get my own funding. Imagine my shock to be told by some of the erudite professors that I had conducted my field research all wrong. I should have given the chimpanzees numbers, not names. And I could not talk about them having personalities, minds capable of rational thinking, and certainly not having emotions because all these characteristics, I was told, were unique to the *human.* The difference between us and other animals, it was maintained, was one of kind, not degree. Fortunately I had had a wonderful teacher during my childhood who had taught me that the professors were wrong—my dog, Rusty! And eventually science came to realize that we are not so unique after all. We are part of, not separated from, the rest of the animal kingdom.

Today many field studies of chimpanzees have been conducted in different parts of Africa, and also of gorillas, bonobos, and orangutans. All of these animals commonly show behavior that reminds us of some of our own. There is their nonverbal communication—all of them kiss, embrace, hold hands, bow, and beg for food (or reassurance) with their palms up. They show emotions that may be similar to—or even the same as—those we call joy and sadness, fear and despair, anger and frustration. Chimpanzees are territorial. The adult males of a community patrol the boundaries of their territory and, if they encounter individuals from a neighboring social group, they may chase and attack, leaving the unfortunate victims to die of wounds inflicted. However, again like us, all the apes show compassion and even true altruism—as when an unrelated individual adopts an infant whose mother has died and who has no older brother or sister to care for it.

All the apes can make and use tools. Moreover, different communities of apes use different objects for different purposes, indicating the behavior is learned: infants and juveniles watch closely and imitate the actions of adults. Since one definition of culture in humans is behavior that is passed from one generation to the next through observational learning, we can say that the apes have their own primitive cultures. We continue to learn more and more about the intellectual abilities of the great apes—they can tackle complex tasks on computers and learn several hundred signs of American Sign Language (ASL). They can recognize themselves in mirrors. They can understand the wants and needs of others and respond appropriately. They have a theory of mind. Some captive individuals love to paint, and those knowing ASL may spontaneously inform you of what their paintings represent.

It is because of these humanlike qualities that the great apes, particularly chimpanzees and orangutans, have been sold as pets, taken into homes, dressed up, and treated as children. They are charming and entertaining—until, around age six or seven, they grow out of infancy and no longer want to behave like human children. At this point, stronger than the humans who own them and potentially dangerous, these unfortunate "pets" may be confined to a cage, tied up on a chain, sold to a bad zoo or circus or, until very recently, used in medical research.

It is well known today that the training of young apes for entertainment typically involves harsh punishment. They will learn through reward systems, but in show biz the trainer wants instant obedience, and this is most speedily obtained through fear. The lives of ex-performers are similar to those of ex-pets.

In the composition of our DNA, we, chimpanzees, and bonobos differ by only just over 1 percent. In fact, the geneticists tell us that the main difference is in the expression of genes that can be influenced by the environment. There are also remarkable similarities in the composition of our blood, immune system, and anatomy of the brain. Gorillas and orangutans are only slightly less different in their biology.

Because of these striking similarities in biology and physiology, many chimpanzees and some bonobos have been extensively used in psychological and medical research. Physical contact with the mother is very important for infant apes. Some infant chimpanzees were taken from their mothers and kept in isolation to produce models that resembled depression in human infants. Others, confined in metal cages only five by five feet, were infected with diseases such as polio, hepatitis, and HIV-AIDS that other animals, less like us, cannot be infected with. The hope, of course, was to find cures and vaccines that would benefit humans. These experiments typically inflicted a great deal of suffering, both physically and psychologically. Nor did they result in medical breakthroughs. In some cases treatments that benefited the apes proved harmful to human patients. And in other cases human trials for certain treatments that did eventually prove to be beneficial for humans were delayed because the medications were harmful to the ape subjects. Moreover, a number of chimpanzees took part in some of the first flights into space. The rigorous training undergone by these youngsters was brutally rigorous—one of the first human astronauts told me that he could not have endured training of that sort.

Gradually, with increased understanding of the true nature of the great apes (and other animals), more and more people—including scientists, lawyers, and activists—have managed to prevent much of this abuse. Conditions in zoos have improved and continue to do so. The use of chimpanzees in entertainment has been banned in many countries. And recently the U.S. National Institutes of Health ordered that all of its chimpanzee subjects (over three hundred) should be retired to sanctuaries after a team of experts found that *none* of the current experiments were useful or even potentially useful for the benefit of human patients.

At the same time, however, the great apes are becoming increasingly threatened in the wild due to habitat loss, the bushmeat trade, and the continuing live-animal trade—mothers are killed so their infants can be stolen and sold. Orangutans are losing vast areas of rain forest habitat due to palm oil plantations and forest fires. Many NGOs are working to care for more and more orphaned apes in sanctuaries, but such efforts can never ultimately be successful in protecting the species unless sufficient areas of their habitat can be protected.

That is why Chris Herzfeld's book *The Great Apes* is so important. To fund protection efforts in the wild and to prevent further exploitation of captive apes, it is necessary to raise awareness about the problems these animals face. It is vital to increase the general public's understanding about the nature and the plight of our closest living relatives before it is too late.

JANE GOODALL, PhD, DBE
Founder of the Jane Goodall Institute & UN Messenger of Peace

Preface

FOR THOUSANDS OF YEARS, HUMANS and apes have cultivated a long history in common. They have coevolved from time immemorial in African or Indonesian tropical forests. They have coinhabited various places since antiquity. So extraordinarily close to humans in their physical and behavioral aspects, the great apes have occupied a special place in the Western imagination since their arrival in Europe in the seventeenth century. The twentieth century was marked by familiar personalities: Cheeta, the companion of Tarzan the ape-man; the mighty King Kong; the mischievous Curious George; the chimpanzees studied and described in *National Geographic* by Jane Goodall (David Greybeard, Flo, Fifi, and Goliath); the gorillas studied by Dian Fossey, later portrayed in *Gorillas in the Mist* (Digit, Peanuts, Icarus, and Pantsy); even characters from *Planet of the Apes,* such as Dr. Zira, Cornelius, Cesar, Koba, and Maurice. Other personalities have appeared more recently in the public eye, further highlighting this singular history between the West and the great apes: the chimpanzee Washoe, the gorilla Koko, the orangutan Chantek, and the bonobo Kanzi, all four "talking apes" that have become media stars; one of the most famous mascots in the NBA, the "Michael Jordan of Mascots," the Phoenix Suns Gorilla; the stubborn gorilla in the famous video game *Donkey Kong;* the chimpanzee Willy, the gorilla Hugh, or the title character of *Gorilla,* all written and illustrated by Anthony Browne.

Out of this centuries-long history with our closest cousins, I did not want to shine the light only on a few primates that have become stars. It is important to understand how humans have perceived apes and monkeys and interacted with them throughout the years, to map out the routes and arenas that brought them together, to examine the manner in which they became variously the objects of learned discourse, laboratory rats, and pets. I also wanted to seriously consider one of their most spectacular characteristics, their capacity for imitation. Their very name in different languages—*ape, singe, Affe, scimmia*—has inspired terms to describe this ability—*to ape, singer, nachaffen, scimmiottare.* Though this aptitude is often mocked, it has, from antiquity to the present, captivated all those who have spent time with primates. It has reached a remarkable degree among those species phylogenetically closest to humans: bonobos (*Pan paniscus*), chimpanzees (*Pan troglodytes*), gorillas (*Gorilla gorilla*), and orangutans (*Pongo pygmaeus*), all classed within the same superfamily as humans, *Hominoidea.* Imitation is an essential means of learning for the great apes.

This capacity for imitation leads them down a surprising road when they live among humans, that of "becoming-human," a notion that echoes that of "becoming-animal" proposed by Deleuze and Guattari in *A Thousand Plateaus.* Young apes that grow up in human worlds appropriate, with remarkable facility, the habits, *savoir-faire* (this term is used throughout the book in the French sense of knowing how to resolve practical problems), and way of life of humans. They even go so far as to integrate the "way of being in the world," the features of the ethos that in principle belongs only to humans. Thus, some apes make signs like they are writing, exhibit modesty, comb their hair, create paintings, admire sunsets, go swimming, and "declare" themselves to be human. Exceptional in both its breadth and depth, this becoming-human is actually a symptom of something much more profound than it first seems. It shows common dispositions and an essential community tied to a phylogenetic chain, millions of years old, between humans, bonobos, chimpanzees, gorillas, and orangutans. From these ties emerge proximities that shape both their present existence and potential evolution: a long infancy, social learning through observation and imitation, the possibility of expressing emotions, different forms of non-verbal communication, the capacity for language, sophisticated socialities, cognitive abilities and technical aptitude, the transmission of traditions, and a wide range of behavioral flexibility.

Thinking of the human species as being unique in the universe and radically separated from other animal species, people in Western societies have often felt threatened by these remarkable physical and behavioral resemblances. When the similarities are too strong, they implicitly carry with them the risk of possible corruption of humans by beasts. Humans responded to this threat with a compulsive determination to accentuate the differences and define the criteria of distinction. This determination coexisted, paradoxically, with the desire to recover a state of original purity, to renew ties with an Edenic world where all species lived in harmony, and to break free of the weight of an ontological isolation, thanks to our closeness with the great apes. Our phylogenetic cousins incarnated this schizophrenia, and thus aroused contradictory feelings.

Throughout the centuries, our relations with the great apes have been characterized by a constant oscillation between attraction and repulsion, between the recognition of the uncanniness of similitude and the experience of the troubling familiarity of the other. Punctuated by this oscillation, the stories and collective representations of the great apes have varied according to the time period. They reveal the ways in which humans defined themselves in relation to the animal as a function of the Zeitgeist. Figures of the devil in the Middle Ages, strange hybrids in the seventeenth century, and objects of study that showed resemblances to humans in the eighteenth century, apes became particularly threatening in the nineteenth century. The repulsion people felt toward them was reinforced by the affirmation, thought to be shameful by many, of a common origin of humans and great apes. Even though they had been classified among the primates since the eighteenth century, humans could not accept having apes as ancestors.

This refutation of the Darwinian hypothesis continued into the twentieth century. It was at the center of a trial against a teacher in Dayton, John Thomas Scopes, who was accused of violating the law forbidding the teaching of the theory of evolution under Tennessee's Butler Act. Scopes had taught the theory of evolution to his students and thus denied the divine origin of humans. The Scopes "Monkey Trial" took place in 1925, at the same time that systematic studies of primates were just under way. In part fed by the results of this research, new narratives gradually modified people's views of our closest cousins. I attempt to understand what allowed this change of view by examining the pivotal changes that occurred in the construction of knowledge about primates throughout history. It seems to me useful to clarify the actual knowledge and practices by returning to this history through an analysis of stories and conceptions about primates from the past to the present day. My objective is not to undertake an exhaustive account of this history, which spans thousands of years, but rather to show the main ideas, the major personalities, and the crucial time periods, at the risk of sometimes simplifying certain debates.

This work encompasses a number of different fields: it is a history of science mixed with philosophy and with some insights gained in visits to places in Europe, Africa, and the United States. Its scope ranges from ancient philosophers to the encyclopedists of the Middle Ages to scholars of the Renaissance to classifiers of the Enlightenment (chapter 1); from museum collections to colonial expansion (chapter 2); from experimental biology to the theory of the mind (chapters 3 and 4); from research in the field to culture among animals (chapters 6 and 7); and from the entry of women to the field of primatology to their important contributions to its renewal (chapter 8).

In addition, special attention is given to the great apes that have lived alongside humans (chapter 5). While I do not condone bringing great apes into peoples' homes, it is nevertheless interesting to take into account these experiences in order to illuminate and elaborate a central concern: what does it mean to be an ape? Expressing a different ethos as a function of time and place, great apes are never quite what we expect them to be, whether as specialists in the botany of tropical forests, communicating with humans using sign language, trying to makes lives for themselves in zoos by becoming interested in our ways of doing things, or presenting themselves as socially sophisticated creatures. Inventive, clever strategists, they never cease to surprise us. It is what makes them so interesting. Thus, rather than focusing on questions proper to humans and a logic of difference, this brief history places the accent on the similarities and closeness between humans and apes as well as on the essential community that unites them.

Acknowledgments

THROUGHOUT *THE GREAT APES: A Short History,* I have tried to follow the great apes in their wanderings (often forced) in our world. My objective was to reassemble the personalities and key periods of the long history that unites humans and apes in a brief work that would be both accessible to the general public and of interest to specialists in primates. I have had the opportunity to meet some of these great apes and would like to pay a special tribute to Victoria. She belonged to a relatively rare subspecies of eastern gorilla called Grauer's gorilla, only two of which remain in captivity. Victoria died on May 21, 2016, at the Antwerp Zoo. My thoughts also go out to the orangutans Solok, Tubo, and Dayu (all deceased), Wattana, Tuan, Nénette, and Chantek, as well as the chimpanzee Panzee and the bonobo Tshilomba.

Many people have accompanied me on this adventure. I will never be able to thank three of them enough for their warmth and support: Jean Thomson Black, Eleanor Sterling, and Kevin Frey. Jean believed in this work from the outset and pushed for its publication. Eleanor's enthusiasm for the book and her advice about this entirely revised version were priceless. Kevin translated the book with brio and nuance while showing great patience and kindness. I would also like to thank the entire team at Yale University Press, particularly Samantha Ostrowski and Michael Deneen. Further, the French edition of this book would never have appeared without Christophe Bonneuil, Dominique Pestre, Olivier Bétourné, Jean-Marc Lévy-Leblond, and Sophie Lhuillier at Editions du Seuil. I wish to express once again all my gratitude for their efforts.

I would also like to thank, with all my heart, Jane Goodall for her kindness as well as Frans de Waal, Ian Tattersall, and Gregory Radick for the confidence they showed in one of my last publications. Many researchers have also graciously helped me during the revision of this work: thanks to Jacques Vauclair, Eugene Linden, Patricia Van Schuylenbergh, Jacques Cuisin, Jacques Pasteels, Eric Baratay, Tom Baione, and Richard

Burkhardt. I would also like to mention several of the people passionate about great apes whom I met during my travels around Africa, Europe, and the United States who had the kindness to share their experiences with me: Lyn Miles, Gen'ichi Idani, Claudine André, Patti Ragan and the team of the Center for Great Apes, as well as the staff at the sanctuary Save the Chimps, with a special recognition of the life-saving work of Carole Noon. I also want express my gratitude to the director of the State Darwin Museum, Anna Kliukina, as well as Anna Kurzaeva and Xenia Bogza. And I cannot forget my accomplices at the Ménagerie of the Jardin des Plantes in Paris: Michel Saint Jalme, Jacques Rigoulet, Christelle Hano, and Gérard Dousseau. Finally, I would like to give a warm thanks to several faithful friends: Henry Raven, Anne-Marie Roviello, Dominique Lestel, Marion Thomas, and Lidia Breda. I will end by expressing my deepest gratitude to those dearest to my heart, Henry, Lauren, Nathan, and Elisabeth Herzfeld.

The Great Apes

Le Jocko et le Pongo, illustration by Antoine Chazal (1793–1854). Buffon wrote: "We present these two animals together, because they surely must constitute a single species."
(From Buffon, *Œuvres complètes*)

ONE

The Uncanniness of Similitude

Wild Men, Simians, and Hybrid Beings

WE SHARE A LONG HISTORY WITH primates. Since antiquity, humans have been fascinated by their remarkable resemblance to us as well as their extraordinary talent for imitating our behaviors. Numerous works of art bear witness to this capacity to "ape" humans. However, after the time when hybrid gods, half human, half animal, were worshipped, people began to fear the forbidden mixture of opposite natures, human and animal. Aversion became juxtaposed with fascination. Primates both captivated and repelled since they showed themselves as similar to humans while exhibiting all the characteristics of savagery and bestiality. In order to guard against any confusion and to defend the "majesty of Man," primates were turned into incarnations of ugliness and vileness, figures of the devil, a mark of sin, lewd and grotesque puppets, or caricatures of humans, depending on the time period.

The first chimpanzees arrived in Europe in the seventeenth century. They remained extremely rare up until the beginning of the nineteenth century. Throughout this period, scholars set out to describe precisely the morphological and anatomical similarities and dissimilarities between humans and primates, thus turning both of them into objects of knowledge. From the eighteenth century onward, they tried to gather together all available knowledge about primates, while being careful to maintain their insuperable differences with humans. Primates were assigned a specific role, that of "a being of nature," a mute and savage figure governed by instinct and the needs of the moment. Created in God's image, spiritual, endowed with reason and language, humans were placed at the peak of the great chain of being: masters and rulers of nature. In conjunction with a compulsive will to determine the criteria of distinction between these related species, there was also a constantly renewed sense of curiosity and astonishment.

Humanized Primates and Monkey Orchestras

Numerous artistic representations of primates have been found in the Mediterranean: Mesopotamia, Egypt, Crete, Greece, and various countries in the Middle East. Surprisingly, although emblematic of their talent as imitators, the primates are represented dancing, wearing rich clothing, playing the harp or the lyre. In ancient Egypt, the flow of commerce and the exchange of prestigious gifts among the powerful favored the introduction of exotic species such as monkeys, primarily baboons and colobus monkeys. The most often represented primate was the hamadryas baboon, which appeared sitting on wagons full of goods, clinging to the neck of giraffes, kept on a leash, or sitting on a merchant's shoulder. Kept as pets, monkeys lived in the private apartments of the elite. They were depicted attached to the side of a throne, a piece of fruit in their hands. They were also shown dancing or as musicians playing the mandolin, flute, or lyre. These representations appeared in Mesopotamia beginning in the second millennium BCE. One of them depicts a monkey serving drinks to an orchestra made up of various species of animals. In Babylon, they were honored during religious rites or adopted by private owners. The Assyrian king Asurnasirpal II exhibited them among the remarkable species kept in his parks.

These different examples show that from ancient times onward, humans have been captivated by primates' capacities for imitation, and they have brought these creatures into their daily lives. Whether in Asia, Palestine, Syria, or islands in the Aegean Sea, statuettes, pottery, amulets, and vases attest to their presence. They were obviously kept quite close to their owners, carried on their shoulders or perched atop their heads, and sometimes they appeared richly dressed or wearing necklaces. More intimate scenes of monkeys showed mothers holding their babies in their arms.

Primates are also mentioned in the Bible, included in the lists of merchandise imported by the Hebrews. Their strong physical and behavioral resemblance to humans made them, however, dangerous in the eyes of scholars at the time. Seeing them in a dream, for example, was interpreted as a bad omen. They were considered inferior to humans, this inferiority revealed by their ugliness. Primates were to Adam what Adam was to God. Although ridiculed, they were nonetheless seen as objects of luxury. It was prestigious to own one. In ancient rabbinical tales, they fulfilled the functions of servants. They were taught, for example, to pour water over the hands of their master and his guests.

The Domestication of Primates

Phoenician merchants who had bases along the northern coast of West Africa, where monkeys were kept as family pets, greatly contributed to the introduction of African primates to countries where they had been unknown. Few live monkeys were brought in during the Minoan-Mycenaean period (2700–1200 BCE). Only the wealthy imported them or received them as gifts. But primates were very highly prized and gradually became more numerous. Phoenician merchants also facilitated the diffusion of Egyptian, Assyrian, and Eastern artistic themes in the Aegean, then the Mediterranean. Representations of primates dating from 1400 BCE have been found at various sites.

One of the most beautiful illustrations comes from Crete, at the House of Frescoes at Knossos (late Minoan): primates colored in blue, represented in their natural environment, and undoubtedly drawn from nature. From this period, frescoes, vases, and statuettes prove that there existed spaces of cohabitation between humans and different species, including baboons, in the Aegean. The exchanges established between Greece and the Aegean islands, Mesopotamia, and Egypt continued to influence the diffusion of representations of primates in the ancient Greek world. Imported by the Carthaginians and the Phoenicians, who had lived alongside the species for some time, Barbary macaques became fairly common in Greece, where they were particularly appreciated as pets. They also appeared in artistic themes within other Mediterranean civilizations, notably the Etruscan.

However, all of these encounters were with monkeys rather than apes. One of the first Westerners to see a great ape in its natural environment was a Carthaginian admiral named Hannon. Around 500 BCE he was exploring the coast of West Africa with a view toward colonization. Three days after his departure, he landed on an island in the middle of a luxurious bay, somewhere near present-day Gabon. He found there a "savage people," covered in hair, that his interpreters called Gorilla. His soldiers captured three females of the troop. Attacked by the primates, the soldiers killed them in self-defense, then skinned them. Hannon brought the trophies back to Carthage. Later, Pliny the Elder noted the presence of these "Gorgons" in the temple of Juno in 146 BCE, the year the city was destroyed, after a Roman siege that ended the Punic Wars. Were they pygmies, chimpanzees, gorillas, baboons, or a species of ape that has since disappeared? Many believe this is the first mention of a gorilla by a Westerner. The "domestication" of these gorillas was limited to the exhibition of their remains. It would not be until the twentieth century that an ape of this species would be adopted by a human family, whereas monkeys have been kept as pets since antiquity.

The Most Beautiful Ape Is Ugly, Compared to Man

Since the seventh century BCE, primates have appeared in poems, fables, satires, and philosophical writings. They have incarnated ugliness and the most extreme disgrace or figured as maleficent beings, manipulative and cunning. In the sixth century BCE, the date from which primates become more familiar to the West, they played different roles in the fables of Aesop, especially that of the antihero. During the classical period, they were considered the most hideous and shameful of beasts because they tried to pass for human. They were nothing more than arrant imposters and vile imitators. According to Plato, monkeys (*pithêkos*) were endowed with intellect without depth and thought without sense. The philosopher said, "The most beautiful of monkeys is ugly compared with the race of man," adding that "the wisest of men, if compared with a god, will appear a monkey, both in wisdom and in beauty and in everything else."[1] Such sentiments would long stain the reputation of primates.

In his *History of Animals*, Aristotle writes about the dual nature of primates. In his discussion of primates (including Barbary macaques, baboons, and other tailed monkeys) in part 8 of book 2, he says that they "share the properties of man and the quadrupeds." He adds that the face of such creatures "resembles that of man in many respects; in other words, it has similar nostrils and ears, and teeth like those of man, both front teeth and molars." Primates also have hands, fingers, and nails, "only that all these parts are somewhat more beast-like in appearance," and they use their feet as if they were hands.[2]

Many other authors, however, described primates in a positive fashion. In North Africa, some families kept Barbary macaques. They were educated, performed various chores, and became playmates for the children. Some even ate at the table with their human companions. Desmond and Ramona Morris relate that "the simians were deeply respected by the inhabitants who gave their children ape names. The penalty for killing a monkey was death."[3] This positive image was not shared by Cicero who, quoting Quintus Ennius, declared, "How like us is that ugly beast, the monkey!"[4] Become proverbial by force of repetition, this sentence sums up the *simia* in a characterization that would successfully cross the ages, up until the beginning of the twentieth century: this ugliness renders the resemblance to humans problematic.[5]

Calling primates "exotic imitators of humans," Pliny the Elder, the author of *Natural History*, evoked their startling similarity to humans and praised their extraordinary intelligence.[6] Gathering all the knowledge of his time concerning simians and hybrids, he stated that there were three species of primates: Choromandae, Syrictae, and Satyres (baboons, orangutans, and gibbons). Later used in the binomial Linnaean nomenclature as an adjective (*satyrus*), the term *Satyre* refers to the *sylvan satyrs* of the god Dionysus in Greek mythology—hairy, two-legged creatures possessing a libidinous nature. The Greek historian Plutarch thought that primates could serve no other function than that of amusing the public. In the second century CE, the sophist Aelien also attributed human behavior to monkeys. He described them as coachmen, able to handle, using a whip, a team of goats. At this time the Romans were familiar with different species of primates.

In the Roman Empire, one of the finest physicians of antiquity, Claudius Galenus, recommended that his students dissect primates in order to better understand human anatomy. (The study of human cadavers was forbidden in Rome in the second century.) Based on his dissections, Galenus was the first to affirm a morphological and anatomical proximity between humans and primates. He remained the authority on this topic until the middle of the sixteenth century, when Andreas Vesalius questioned the accuracy of his studies of human anatomy. As far as primates were concerned, however, his work was deemed "precious and valuable" by the Dutch anatomist Petrus Camper toward the end of the eighteenth century.[7] From his examination of Galenus's accounts, Camper (as had Tyson before him) believed the Greek physician had dissected chimpanzees (or orangutans), gibbons, and baboons.

Toward the middle of the third century, Gaius Julius Solinus listed five species of primate: the cercopitheci, the cynocephali (many of which lived in Ethiopia and had faces similar to a dog's), the sphinges (easily domesticated), the satyri (beings with goat legs, related to centaurs), and the callitriches (who had bearded faces). Solinus was so faithful to Pliny the Elder that he was often called "Pliny's monkey." The gibbon would long be called "Solinus's Satyre." Read by many, his works on the marvels and curiosities of the world would play an essential role in the constitution of zoological knowledge in the Middle Ages. The different species of primates named by the ancient authors would figure, moreover, at the center of variations of the theme of the wild man (or *Homo sylvestris*) until the start of the twentieth century.

Primates as Figures of the Devil

From the fall of the Roman Empire to the Gothic era, the medieval world was peopled with beings that oscillated between the animal and the human, threatening the boundaries erected between the two realms. Scholars compiled knowledge from the church elders and the writings of the ancients:

naturalists, philosophers, and physicians. Focused on humans, monsters, and animals, books 11 and 12 of the *Etymologiae* compiled by Isidore of Seville in the seventh century would greatly influence medieval bestiaries. The work enjoyed renown up until the Renaissance.

The peasantry of the Middle Ages maintained a pagan imaginary, continuing to worship, despite the efforts of the Catholic Church, figures tied to cults of nature and fertility. Ancient belief systems, particularly the Egyptian religion, whose pantheon is full of hybrid and animal deities, were considered to be incompatible with Christian doctrine. The divinity attributed to the baboon by the Egyptians, a people who worshipped idols, could not fail to elicit disapproval, as did other pagan cults. Christian authors also radically differentiated the descendants of Adam from the monstrous races of the terrae incognitae described by the ancients: icthiophagi, cynocephali, panotii, troglodytes, and other acephali, which included the "simians."

Associated with profane beliefs, primates posed a problem for theologians. They could not be seen as the result of a divine failure, since God created all beings as perfect. They were not really demons, as were centaurs or cyclops. How, then, to think about these beings dangerously similar to humans? Combating pagan and polytheistic practices and exerting a strong influence throughout the Middle Ages, the *Physiologus,* a Christian bestiary compiled between the second and fourth centuries CE, was an inventory of animal species, real and imaginary (such as the unicorn or griffon). It presented primates as being one of two species, along with the "wild ass," that personified the devil.

The "ape-devil" theory would hold sway throughout the Middle Ages. Exhibiting an ugliness that symbolized evil and revealed their bestiality, primates attempted to imitate humans, just as the devil, a hybrid being par excellence, tried to imitate God. Transformed into religious symbols, they were the prototype of the trickster figure. If humans were fallen angels, primates were humans degraded by a divine act, deluded human sinners. Personifying the risk of degeneration and a regression toward animality, they served as warnings. Throughout the Middle Ages the Catholic Church considered monkeys as figures of the devil, *figura diaboli,* in theological and moral discourse as well as in traditional iconography. It appears, however, that despite its widespread diffusion, the doctrine of the simia being related to Satan had little effect on popular conceptions of primates, perhaps because of the rarity of official images of the devil adopting this form. The Evil One was more usually incarnated by other animals such as the dog, cat, or goat.

Between the eleventh and twelfth centuries CE, Barbary macaques arrived in Europe with their Eastern trainers, who wanted to develop an entertainment market for them within the economy of the period following the Crusades. Up until the fourteenth century, they were the only primate species well represented in developing medieval cities. The macaques amused crowds and were sold as prestigious items. Primates, having become more familiar, were also objects of direct observation. Thus, it would not have been very believable to maintain them in their role as devils incarnate. Toward the end of the Middle Ages, other traits were projected onto them: frivolity, vanity, gluttony, and obscenity as well as folly and impulsiveness. Through these characteristics, primates continued to symbolize sin, constituting a warning against the fall of humans, struck down when they repudiated their spiritual dimensions.

Monkeys, Apes, and the Encyclopedists of the Middle Ages

The authors of Christian bestiaries strove to gather together all knowledge of animals, from scientific texts to artistic representations, from moral allegories to traditional adages. Animals became authoritative examples upon which authors fixed their moral precepts. Various encyclopedists continued to assert the relationship of monkeys and apes with the devil up until the fourteenth century. In the twelfth century, scholars began to look back to ancient sources of information, rediscovering Greek and Latin authors. Interest in hybrid beings, fauns, apes, monsters, and various mythological creatures resurfaced. The physical resemblances between humans and primates were recognized, but their commonality was neutralized because of radical differences: the soul and reason belonged to humans alone.

The medieval encyclopedists renewed studies of primates and became interested in their "mentality," even their "psychology." Furthermore, the relationship to received knowledge from books began to change. The pioneering work of Roger Bacon advocated the study of nature through empirical methods. The author of the monumental *De Animalibus,* Albertus Magnus, developed his natural history as a branch of theology, as did other encyclopedists. His work nonetheless differed from theirs. A great scholar of Aristotle, he also based his descriptions on direct observation. He identified all the similarities between humans and other primates, including mental capacity (memory, judgment, and imagination). He believed, however, that reason was a fundamental characteristic of humans alone—only they could make laws for themselves, differentiate between good and evil, and live in civilized societies. The Dominican theologian Thomas of Cantimpré believed that the physical similarities between humans and primates were misleading. He considered simian imitations as the worst copies of human behavior. He added that since monkeys went about on four limbs, by nature they were forced to turn their gaze to the ground, whereas man looked toward the heavens, the source of his salvation.

Strange Hybrids

The Renaissance was marked by a recovery of ancient heritage and the compilations of the medieval encyclopedists. The scholarly world was thus haunted by accounts of hybrids from the cultures of Greece and the Middle East; these creatures, according to Pliny and Solinus, lived in far-off countries such as India or Ethiopia. The great apes were from the outset associated with these beings at the edges of the world, half human, half animal, who, in the imagination of the time, were part of the outlying enclaves inhabited by a degraded humanity, bestial and mutant, consisting of fauns, strange gods, wild men, and monsters. The illustrations that accompany the descriptions of these anthropomorphic beings reveal their fictional character.

Bernhard von Breydenbach, archbishop and dean of Mainz, offered a single zoological engraving in his *Travel in the Holy Land* (1486). The representations of animals Breydenbach claimed to have encountered during his pilgrimage in the Holy Land include giraffes, crocodiles, long-eared Indian goats, camels, salamanders, and unicorns. In the lower right-hand corner of the engraving, an anthropomorph with a tail is holding the leash of a camel and leaning on a crutch. In between a human and an ape, it cannot move like a human without this aid. Reproduced by a number of naturalists (Gesner, Aldrovandi, Tyson, Boreman, Linnaeus, Haag, Buffon, Vosmaer, Edwards, Hoppius, Haeckel, and others), this representation of an anthropoid leaning on a cane, associated with the great apes, would remain firmly rooted in the Western imagination. It can still be seen today, for example, in an exhibit in the Hall of Primates at the American Museum of Natural History, where there is a skeleton of a young ape standing up, its hand on a cane. It is shown alongside a young *Homo sapiens* posed in the same attitude. Like the ape, the human infant has not yet attained the perfection of adult humans, inscribed in their bipedal means of locomotion.

Voyages of Exploration and the Inventory of Nature

Beginning in the fifteenth century, European sovereigns and merchants financed the immense enterprise of the great voyages of exploration. Though they were initially undertaken for economic reasons, these voyages also allowed the collection of a wealth of knowledge in different domains, including art, geography, anthropology, and the natural sciences. Voyagers' tales completed the written knowledge on primates, notably the *Historiae Animalium* of the Swiss naturalist Conrad Gessner (considered the pioneer of modern zoology) in the middle of the sixteenth century. During the same period, the celebrated anatomist Andreas Vesalius was also interested in primates. He showed that the anatomical descriptions of humans made by Galenus in second-century Rome actually were of primates. He confirmed his hypothesis by publicly dissecting a human and a primate in Bologna.

Thus, thanks to voyagers' tales and the work of several well-known scholars, knowledge about primates slowly began to accumulate. Several other accounts from travelers would add still more. A Portuguese traveler, Valentin Ferdinand, participated in a voyage of exploration of the African coast at the turn of the sixteenth century. During his stay in Sierra Leone, he saw anthropomorphic apes, undoubtedly chimpanzees. In 1591 Filippo Pigafetta transcribed the accounts of the Portuguese navigator Duarte Lopez in his description of the Congo. Lopez alludes to various primates, including the chimpanzee. He stated that these animals greatly resembled humans and that they could imitate human behavior to a remarkable degree.

Concerning gorillas, one of the oldest descriptions, after that of Hannon, is by an English corsair, Andrew Battell, published in 1613. Battell spent a number of years in Portuguese Angola and Congo. He talked of two kinds of "very dangerous monsters" feared by the local population. The biggest of these, called *m'pungu* in the local language (*pongo* to the Europeans), was undoubtedly the gorilla. It resembled humans in its proportions but was much bigger and covered in hair. Battell added that it was sociable and had a humanlike face. He believed the primate walked upright on two legs. He spoke of pongos sleeping in trees, in nests that they built themselves, and constructing a kind of roof to protect themselves from the rain. He also observed a smaller primate than the pongo: the *m'geko* (translated as *engecko* or *enjoko*), probably the chimpanzee. Despite Battell's account, the gorilla would not appear in scholarly texts until the end of the eighteenth century.

Jarric's Chimpanzee and Bontius's Orangutan

Sent to Africa toward the end of the sixteenth century, the French Jesuit Pierre Du Jarric described different species of primates from Guinea. He reported that the *barri* (probably a chimpanzee) easily learned human ways of doing things and could be trained to perform certain tasks: it would, for example, go fetch water from the river. Samuel Blommaert, from the Netherlands, who stayed in Sambas (Borneo) in 1609 and 1610, was told by the local rulers that large apes lived in the area. King Timbang Paseban told him that these primates, attracted by human women, would carry them off and rape them. This behavior would long be associated with orangutans.

Another physician from the Netherlands, Jacob de Bondt, called Bontius, also had the occasion to see the red ape. Sent as chief doctor by the

Dutch East India Company to Batavia (now Jakarta, in Java), he explored the region for nearly four years, observing gibbons and orangutans. He spread the Javanese expression "man of the woods" (*Homo sylvestris*), which reflected the intermingling of wild characteristics and human traits. Discussing the local belief that these apes were the fruit of an impure commerce between apes and Indian women, Bontius saw only primitive credulity. He also described the behavior of a female orangutan that adopted certain human habits: she cried and exhibited modesty. The only thing missing was speech. As to this, the doctor reported that "the Javanese claimed that the *Orang-Outangs* could talk, but that they did not want to because they did not want be forced to work: ridiculous, by Hercules."[8] Bontius clearly did not believe this explanation. The capacity for language constituted an important criterion of distinction between humans and the great apes. In 1661 Samuel Pepys thought that chimpanzees were capable of understanding spoken English and could be taught to express themselves by means of signs.

The Arrival of "Satyrs" in Europe

In the seventeenth century, the arrival of great apes on the European continent was one of the most important events in the history of the natural sciences, which were already changing drastically. In his *Novum Organum* (1620), Francis Bacon distinguished his new method for the advancement of science from the more bookish erudition of the ancients. He advocated knowledge based on observation, an empirical approach, and experiments. René Descartes, in his *Discourse on Method,* wrote of an "animal-machine," a speculative vision and mechanistic construction of an animal compared to an automaton. While this thesis functioned with those animals closest to humans, the primates, it could scarcely be applied to others. Scholars were able to apply these empirical methods and Cartesian precepts to the great apes to discover their secrets now that the first specimens had arrived in Europe. However, the majority of the apes died on the voyage, and were thus more often studied as cadavers. They were observed to be remarkably similar to humans. Were they monsters, animals, humans, or hybrids? Renaissance scholars would study them to find out.

Around 1630, a female chimpanzee from Angola was sent as a gift to Frederik Hendrik, Prince of Orange-Nassau, at the Hague, for one of the many menageries the prince owned. It was the first living great ape to arrive in Europe. Believing it to be originally from the tropical mountains of India, the Dutch physician Nicolaes Tulp, or Tulpius, a colleague of Bontius, classified it among the "satyrs" of the ancient writers and named it *Satyrus indicus,* referring to it also as *Homo sylvestris* and *orang-outang,* even though it was actually a chimpanzee.

Copperplate of *Homo sylvestris* or *Orang-outang* (actually a chimpanzee). (From Tulpius, *Observationum Medicarum,* 1641)

This multiplicity of names, added to the imprecision of the illustration, did not do much to help the identification of the two known species of anthropoid primates, the chimpanzee and the orangutan, which were often confused up until the turn of the twentieth century. It is even possible that the ape described by Tulpius was a bonobo. In recent times, the primatologists Vernon Reynolds and Frans de Waal have examined Tulpius's drawing, the first done of the species from a living specimen, which was widely circulated (even though it was not highly realistic). They think that it may well have been a bonobo because of an anatomical detail more common among bonobos than chimpanzees: a cutaneous connection between the second and third toes.

Having the occasion to observe the young female alive, Tulpius explained that she drank by holding a pot in one hand and lifting its cover with the other, then delicately wiping her lips after drinking. When going to bed she covered herself with blankets, as would humans. She died in 1632. A celebrated surgeon and anatomist (painted in 1632 by Rembrandt in *The Anatomy Lesson*), Tulpius published the first scholarly description of a great ape based on behavioral observation and morphological study in his *Observationum Medicarum* (1641). Unable to preserve the cadaver, he did not provide an anatomical description.

The Anatomy Lesson of Professor Tyson

In 1693 an eminent specialist of natural history, John Ray, attempted an initial classification of primates based on their morphological characteristics. During this pioneering period of the advancement of English science, marked as well by Isaac Newton and Robert Boyle, a number of scholarly societies had sprung up, including the Royal Society of London (1660) and, in Paris, the Academy of Science (1666). Descriptive and comparative anatomy became an important concern of scientists. A young male chimpanzee arrived in London from Angola in 1698. He died three months later from an infection.[9] Having already done several important anatomical studies of other species, including a dolphin in 1680, Edward Tyson dissected the chimpanzee in April 1698. He made detailed observations, complementing them with the results of his dissection of many specimens of monkeys.[10]

Tyson's work, the first comparative anatomy study of primates, made a big impression, as did his presentation at the Royal Society of London. His study constituted, along with Ray's advances in the area of primate classification, the most important contribution to the knowledge of apes before the eighteenth century. Carefully noting all the similarities (forty-eight common points) and the differences (thirty-four) between his "pygmie" and two different species of monkeys, then between this same "pygmie" and humans, Tyson showed the troubling parallels to humans (more numerous with the latter than with other primates), their anatomical proximity being, for the first time, demonstrated scientifically and unequivocally.

The Pygmie, drawn by William Cooper and engraved by Michael van der Gucht. (From Edward Tyson, *Orang-Outang, sive Homo sylvestris,* 1699)

Furthermore, Tyson described various behaviors of the chimpanzee he had observed before its death. The young ape slept on a bed, his head on a pillow, with a cover pulled over him. He showed pleasure in getting dressed and readily let others help him when he had problems putting on clothes. Finally, the anatomist made a philological analysis of the terms currently applied to different species of primates (*orang-outang, Homo sylvestris, monkey, ape*) as well as the expressions of the ancient writers (*pygmies, cynocephali, satyrs,* and *sphinges*).

Scholars thus continued to refer to the old half-realist, half-mythical bestiary. Tyson, however, said that these "hybrid beings" were not monsters but monkeys. He concluded that the "pygmie" specimen he studied should be placed in a special group; although it was still an animal and not a hybrid, it was different from the others. He thus maintained the dominant position of humans as radically separated from animals, however close they may be. He also instituted the name "orang-outang" as a generic term widely used for all great apes then known.

While the works of Tulpius (1641) and Tyson (1699) established a close anatomical relationship between humans and great apes, the latter nonetheless continued to be thought of as beneath humans on the great chain of being, the similarities at this time being seen as strictly physical. The "pygmie" had the physical structures and the "equipment" to allow it to think and talk, but it could not, since it lacked a conscious mind. The great apes thus were something in between, an intermediary between rationality and animality, between humans on the "lowest scales" and the highest of primates.

Apes, Natural History, and Classification

In the eighteenth century, the public had an unparalleled passion for natural history. Rare and stunning species such as the rhinoceros shown at the Saint-Germain fair (1749) or the elephant exhibited at Reims (1773) were all the rage. People could see exotic animals thanks to traveling shows at fairs. The imitative abilities of monkeys were put to use during the spectacles. Dressed up, they mimed certain professions, different human activities, and scenes from daily life. They played cards, ate with utensils, or imitated fashionable dances. These monkeys, whose very name evoked in 1827 "an idea of thoughtlessness, indecency, burlesque bestiality," were already being mentioned by some scholars as possible ancestors of humans.[11]

Still influenced by ancient thinkers but increasingly sensitive to modern scientific methods, naturalists pursued work that drew from scholarly studies, observation, and theology. Reports from eyewitnesses remained rare. At the beginning of the eighteenth century, the English captain Daniel Beeckman was among the first to write about the orangutans of Borneo in *A Voyage to and from the Island of Borneo* (1718). Having kept an orangutan for several months, he reliably described different traits and behaviors of the primate, although he was mistaken in saying that the ape walked upright.

Within museums, the arrival of exotic specimens of undescribed species pushed scholars to undertake the Herculean task of making an inventory and classifying all the objects in the world. Within this framework, they privileged morphological and anatomical studies, observation of external characteristics, analysis of the internal structure of animals, and comparative anatomy. Their will to master chaos also expressed itself in the practice of denomination. Scholars found in the Linnaean system the possibility of undertaking this communal task according to rigorous standards of analysis and shared codes, around which they formed a collective. In 1735 Carl Linnaeus managed to systematize a new nomenclature called "binomial," that is to say, composed of two terms: the first designated the genus, the second the species. Bringing the great chain of being into his *Systema Naturae,* he defined the order Anthropomorpha from characteristics common to humans, primates, and sloths.

In 1758 Linnaeus published the tenth edition of *Systema Naturae,* considered as the starting point of zoological nomenclature. He changed the term Anthropomorpha to Primates (derived from the Latin *primas,* "of the highest rank"). Primates then comprised four genera: *Homo, Simia, Lemur,* and *Vespertillo* (bats). In the twelfth edition (1766), the group including humans (*Homo*) was divided into two subgenera: *Homo diurnus,* or "man of the day" (among them *Homo sapiens,* which included *Homo ferus* and *Homo monstrosus,* groups in which "wild" men and "wild" children were classified), and *Homo nocturnus* (or *Homo troglodyte*), which included chimpanzees (*Simia satyrus*) and orangutans (*Homo sylvestris*).

A creationist and fixist, Linnaeus spoke always of God, seeing nature as constituting a reservoir of sensory evidence that attested to divine existence. Thinking of himself as a "second Adam," he declared, "I name species as God created them." He thus obeyed the biblical prescription of denomination, doing so by using "names from mythology generally pleasing to the ear."[12] Linnaeus created a framework that allowed scholars to establish a structured and unified inventory of all the plants and animals in the world. He was, however, strongly criticized, notably by Buffon (in France) and Blumenbach (in Germany), because of his classification of *Homo sapiens* in the same series as animals, next to the orang-outangs.

Chimpanzees Enter the Scene

At this time, no one went to study primates in the field. Only privileged scholars had occasion to observe the rare living specimens that arrived in Europe. A chimpanzee was first presented to the general public in September 1738 at the Tower Zoo in London; this was also the first time the animal was designated by the name we know today (derived from the Bantu word *kimpanzi* or the African term *quimpezes*). The young female ape was put in a scene depicting "teatime." Brought from Angola aboard the ship of captain Henry Flower, "Madame Chimpanzee," twenty-one months old, was dressed in a silk robe. She drank her tea very daintily and slept in a bed. She walked upright, leaning on a cane, conforming to a now-classic representation. Thomas Boreman described her character as modest, sweet, discreet, and tranquil, words he might have used for a genuine lady. She could imitate human behavior to a surprising degree.

The first living chimpanzee brought to France arrived in Paris in 1740. Buffon, well acquainted with the work of Tyson, which he had translated into French, was then working on his immense *Natural History*, which was published in thirty-six volumes (and seven volumes of supplements) between 1749 and 1789. In it he defined the orang outan as walking upright, having no tail, and possessing hands, fingers, and nails that resembled those of a human. The great naturalist stated: "Of all the monkeys these most resemble man, and are, consequently, most worthy of observation."[13] He compared the anthropomorphic apes to men in a state of nature: they had the same size, the same force, the same ardor for females, walked on two feet, and used weapons. He explained furthermore that the chimpanzee that he had observed, the famous Jocko (also probably from Angola), had picked up certain human habits. It was able to serve itself tea and used utensils, napkins, glasses, cups, and a saucer. He believed that this species could display feelings, possessed a memory, and was capable of learning. He also thought, however, that elephants were more intelligent. Also, while apes had a flattened face, like a human, they were only puppets wearing a "mask of a human face."

Apes and Humans

Severely criticizing the Linnaean classification of humans among the primates, Buffon refused to humanize the orangs-outangs: the observed similarities were purely anatomical, not mental. The author of *Natural History* situated the essential difference between primates and humans in perfectibility and reason, the noblest traits of human beings.[14] Only humans had self-knowledge and could express their inner thoughts and feelings, able to gather their reflections and communicate them. They possessed a soul, thought, and language, all three disconnected from the corporeal. While primates were capable of imitating certain human actions, this was only caricature and simulation: primates could "counterfeit," but they were not capable of pursuing a real project by mere copying.

Their capacity to string together voluntary thoughts to achieve a goal guaranteed the preeminence of humans over these species of a purely animal nature. Buffon thus refuted the idea that humans could have passed through a purely animal stage. He stated, however, that the orang-outang was "the first of the apes or the last of men."[15] Defending a principle of evolution of living things, he believed that apes had undergone a degeneration, which explained their beastliness. Buffon concluded that the orang-outang "is only an animal, but a very singular animal, that Man cannot look at without looking at himself, without recognizing himself, without convincing himself that his body is not the most essential part of his nature."[16] And thus apes were placed outside of humanity.

Concerning the nomenclature and classification of primates, naturalists were in disagreement. Study of anthropoid primates has long been muddied by problems of identification. Furthermore, scholars thought at the time that the orang-outangs could move about as easily on two feet as on four. They were fooled by the ability of the great apes to imitate those around them, a disposition that pushed young primates brought into human contact to adopt their way of walking. From studies on their occipital bone and examination of their spinal column, Louis-Jean-Marie Daubenton proved that the apes were quadrupeds. His studies were in fact motivated by research into an objective criterion of distinction between humans and apes, in this case locomotion. Daubenton showed that vertical posture, an attribute that was as important as language in the division between the human and animal, was a fundamental characteristic of humans. For Daubenton, the great apes living in Europe were thus debased and unnatural. Throughout the eighteenth and nineteenth centuries, a big split occurred between scholars. The majority of them refused the Linnaean classification, wherein humans were classified among animals next to the other primates, as an insult to human dignity. Other naturalists, on the other hand, though less numerous, asked themselves whether apes belonged to the larger human family.

The Orangutan of Arnout Vosmaer

Arnout Vosmaer, director of the cabinet of curiosities and the menagerie of Prince William V of Orange-Nassau at the Kleine Loo (the Hague), was the first scholar to carefully observe the behavior of an orangutan and publish a realistic description: physical appearance, behavior, daily activities. He obtained a specimen from Borneo in July 1776, undoubtedly the first to arrive alive in Europe.

Installing the young female in the menagerie next to his apartments, he observed her for seven months, until her death. He reported that she sought out the company of humans, embraced them, walked upright, appreciated wine, and behaved well at table: "She was taught to eat using a spoon and fork. When she was given strawberries on a plate, it was a pleasure to see how she picked them up one by one and brought them to her mouth with the fork, while holding with the other hand or paw the plate or saucer."[17]

She also used a napkin and toothpick, and made herself a pillow by stuffing straw into a piece of cloth: "Having prepared her bed in the usual way, she took a piece of cloth that was near her, spread it out carefully on the floor, put the straw in the middle, and holding the four corners of the cloth up high, brought this packet with a lot of skill to her bed to use as a pillow, and then pulled the blanket over her body."[18] She seemed to observe humans as much as they observed her. She cleaned the shoes of her visitors and untied the most complex knots. She also tried to pick a lock, wiped up her excrement with a rag, and used a nail as a lever. Impressed by these behaviors, so similar to those of humans, Vosmaer believed that she was very different from all other apes.

In light of the talent of chimpanzees and orangutans that, like those observed by Buffon and Vosmaer, drank tea or ate using utensils and napkins, the naturalist Jean-Baptiste Bory de Saint-Vincent declared that once they were educated, the apes might succeed in becoming "civilized." The French anatomist Xavier Bichat was astounded at this extraordinary ability to imitate, combined with a real intelligence. He considered it a behavioral continuity closely tied to the animals' anatomical continuities with humans, especially those concerning the brain and its organization.

Primate Diversity, Human Uniqueness?

In 1755 Jean-Jacques Rousseau published *A Discourse upon the Origin and the Foundation of Inequality Among Men.* In a seminal note concerning the great apes, he makes the orang-outang the principal actor of the state of nature. Rousseau characterizes this state not as an actual stage in the history of hominization but as a thought experiment that allowed him to conceptualize a presocial state of humanity. In this framework, the fact that an orang-outang can figure among the varieties of the human species is a simple hypothesis.

Until this time, the supposedly "candid" stories of explorers had not been very edifying. These naive observers did not have the skill to catalogue appropriately the beings that, in the guise of the "wild man," existed on the outskirts of humanity. Rousseau opened up a speculative space wherein he described the different candidates of a greater humanity, composed of pygmies, hairy men, giants, orang-outangs, and many other creatures, eventually even humans with tails. For Rousseau, there were more differences between humans than there were between humans and apes. The criterion of distinction put forward before then, language, was itself called into question by the appearance of "wild" children who could not speak.

In 1758, George Edwards, a painter and librarian at the Royal College of Physicians in London, wrote and illustrated a text about the "man of the woods," although he mixed up the characteristics of chimpanzees, orangutans, and gibbons. Criticized by Camper, he was among those who, through their inaccurate descriptions and fanciful illustrations, contributed to the confusion of knowledge about the great apes. In 1763, the naturalist Christian Emmanuel Hoppius added to this confusion. Combining mythology, writings of the classical period, and scholarly observations, he described four types of anthropomorphs: the *Pygmaeus* (probably the orangutan, copied from Edwards), the *Satyrus* (the chimpanzee, under which he put the name of Tulpius, who first described the species), the *Lucifer* (totally imaginary and taken from Aldrovandi), and the *Troglodyta* (fictitious and linked with Bontius). He also defended the idea that anthropoid primates were descended from a cross between humans and apes, an idea soon taken up in arguments applied to other races. The Moors, Hottentots, and other African populations would thus be seen as products of hybridization. Buffon also envisaged fertile unions, forced or voluntary, between male great apes and human indigenous females, an obligatory physical bridge between humans and primates. At the time, the fact that they lived in the same place allowed for their close association.

Chimpanzees, Orangutans, and the Grand Pongo of Batavia

In 1766 the Dutch physician and naturalist Petrus Camper tried to put the classification of the great apes into some kind of order. He believed that orangutans came from Borneo, whereas chimpanzees were from Africa. Able to procure a number of specimens thanks to his connections with the Dutch East India Trading Company, he received the specimen of a young orangutan in 1770. Camper would thus be the first naturalist to compare the anatomy of the red ape with that of humans and monkeys (Tyson having been the first to do so with a chimpanzee at the end of the seventeenth century).

Wanting to prove the uniqueness of the human genus, he would also organize the public dissection of three brains: that of an African, a European, and an orangutan. He found that the anatomical differences between the two humans were minor compared to the differences between the two humans and the ape. He believed that these dissimilarities had been minimized by some anatomists, most notably Tyson, whose work he knew well. In 1779 Camper published "Account of the Organs of Speech of the Orang Outang," in which he showed not only that the male orangutans had laryngeal sacs but also that their speech organs were different from those of humans, contrary to the statements of Tyson and Galen. If orangutans did not speak, it was not due to a lack of a conscious mind, but because their larynx was positioned differently from humans, due to their quadruped stance. Camper confirmed that they were not bipeds, as had long been assumed.

It was not until 1783 that Camper had the opportunity to examine the skull of an adult orangutan. Up to that time, naturalists had been able to examine only very young orangutans, the only ones to have been captured and sent to Europe. They also thought that their infancy lasted only a few months. The adult skull was so different from the other specimens that he classified it as a separate species, called *Pongo,* while those of a smaller size were called *Jocko.* Having been able to examine many specimens, alive or dead, in Batavia (today Jakarta), the German naturalist Friedrich von Wurmb had shown in 1780 the existence of a kind of large orangutan, the

"Grand Pongo of Batavia" (in reality an adult-sized orangutan), thanks to the unusual capture of an individual fifty-one inches tall.

Born thirty years after Camper, the German anatomist Johan Friedrich Blumenbach also made a distinction between chimpanzees and orangutans. Clarifying the former nomenclature, he relegated the mythic creatures and the wild men to a historical note. He brought back, however, Battell's pongo (probably the gorilla) and placed it in the first rank of primates, even though according to scholars of the time this pongo was a pure invention of Battell's. Blumenbach also refuted the idea that the great apes walked upright. He criticized the illustrations in which primates were represented upright and added that in their natural environment they were neither bipeds nor really quadrupeds but were above all adapted to moving through the trees.

Like Buffon, Blumenbach also explained that, once educated, orangutans and chimpanzees were capable of adopting human customs. Nevertheless, again like Buffon, he defended a radical separation between humans and primates. He also noted that the sound-producing structures of the great apes were different from those of humans. Finally, he reserved a special order for human beings, that of "Bimana," whereas orangutans and chimpanzees were ranked among the "Quadrumana."

The Return of the Gorilla

At the end of the eighteenth century, the gorilla, neglected since its mention by Battell in the sixteenth century, was brought out of the shadows by a Scots judge, James Burnett, Lord Monboddo, who mentioned three types of African apes, the largest of which, called *impungu* (a local term from Loango and Congo), would seem to be a gorilla. Describing the primate as extremely large and strong, he noted that its face resembled a human's even more so than did a chimpanzee's.

Having examined an orangutan in France, he declared that the similarities between humans and apes were remarkable on a number of levels: morphological, behavioral, and psychological. Referring to Rousseau, he affirmed a genealogical link between them. Like Julien Offray de La Mettrie, he even believed that the orangs-outans were capable of learning language. He advocated placing them in the human genus, a status some scholars refused to accord to so-called wild people. He admitted that there were differences between humans and primates, but thought that the capacity for speech was not a definitive mark of humanity: the "wild children" (whom he had the chance to meet) had all the appearances of humanity, even though they did not speak, like very young children. Foucher d'Obsonville, who lived in India for twenty years, wrote of a Malaysian guide who perspicaciously described the orangutan as "a wild man of the same genus as humans, though constituting a very distinct species."[19]

The Uncanniness of Similitude, the Troubling Familiarity of the Other

The long history of relations between humans and apes is characterized by a constant oscillation between fascination and repulsion: fascination because of the astonishing similarities and repulsion motivated by the bestial character of this disquieting double. The characteristics the two were seen to hold in common, as well as the criteria of distinction, vary from one age to the next. In the Middle Ages, the Christian world radicalized the rupture by turning primates into figures of the devil. "Aping" human beings while being animals, primates personified trickery. But people were nonetheless amused by their exceptional capacity for imitation.

Placed within an encyclopedic approach that took its coherence from a divine plan perceived in the infinite variety of creation, people during the Renaissance believed primates to be part of an enlarged ontology composed of monsters, wild men, and hybrid beings. Within the dualist framework developed by Plato, Descartes justified the separation between humans and animals by declaring that although their bodies were similar, humans possessed a conscious mind, which led to an insuperable ontological split. The physical and biological continuities that existed between humans and great apes were canceled out by their intellectual and spiritual discontinuity: "Human nature emerged through contrast and discontinuity with animal nature. But it was more about a moral distance than a physical one: the material bodies of humans, bringing them closer to animals, does not account for being human."[20] The defenders of a radical separation between humans and animals would rely for a long time on the concept of the "animal-machine," repeating a simplified version of Cartesian thought.

In the eighteenth century, the anatomical and morphological similarities were confirmed but, in a world still heavily marked by a Christian heritage, they were deemed superficial in the sense that they were strictly structural: apes had not received from God the possibility of having so-called superior functions, especially those tied to reason, language, and morality (in the sense of following rules on how to live in order to save one's soul). Apes thus could pretend neither to the intellectual capacities of humans, created in the divine image, nor to their capacity for perfectibility, human perfectibility being at the core of the definition of humanity, particularly for Buffon. Against the threat of a dissolution of the human species into the series of animals by reason of their classification within the order of Primates in the Linnaean system, it was necessary to maintain the dualist paradigm and reinforce the separation: the revision of the taxonomic status of humans could bring about an irremediable reform of their metaphysical status.

At the end of the century, naturalists discovered the first adult specimens of great apes. Until then, they had studied only very young individuals and had been impressed by their extraordinary capacity to adopt human habits and savoir-faire. Once educated, the young primates showed every sign of civility. It was otherwise for adult primates: they were thought to become more savage and brutal as they got older. Scholars thought that their brains ceased to develop once they reached adulthood, causing them to swing back toward ferocity and savagery. From being civilized, they regressed to monsters, to the "man of the woods."

What were the great apes, then, in the eighteenth century? For most people, they were empty shells of human appearance, empty of anything truly human. Resembling humans but incomplete and mute due to a degradation of their superior faculties, apes were dominated by their muscular activity, their senses, the irrepressible needs of the instant, whereas humans were "self-domesticated," thanks to their mind and intellect. As humans grew older, they became more noble, whereas apes became more bestial, subject to the corporeal and sensory world, which obstructed thought and prevented self-perfection.[21]

Great apes thus provided an extraordinary occasion to think of "ways of being in the world" that preceded the emergence of language, culture, and civilization. Incarnating humans in the brute state, the "primitive," they were used in anthropological and philosophical debates about the state of nature, just as they would be implicated in conjectures tied to the question of the "missing link" in the nineteenth century. These debates were fed by the timely appearance of the "wild" children, who lived, literally as well as figuratively, on the edges of humanity. This was especially true in Europe from the seventeenth century onward, specifically the eighteenth century, as in the case of Victor de l'Aveyron or Marie-Angelique le Blanc. Objects of fascination as creatures occupying intermediary territories between two worlds, wild children and apes constituted new occasions for questioning what it means to be human as well as for exploring the place of humans in nature.

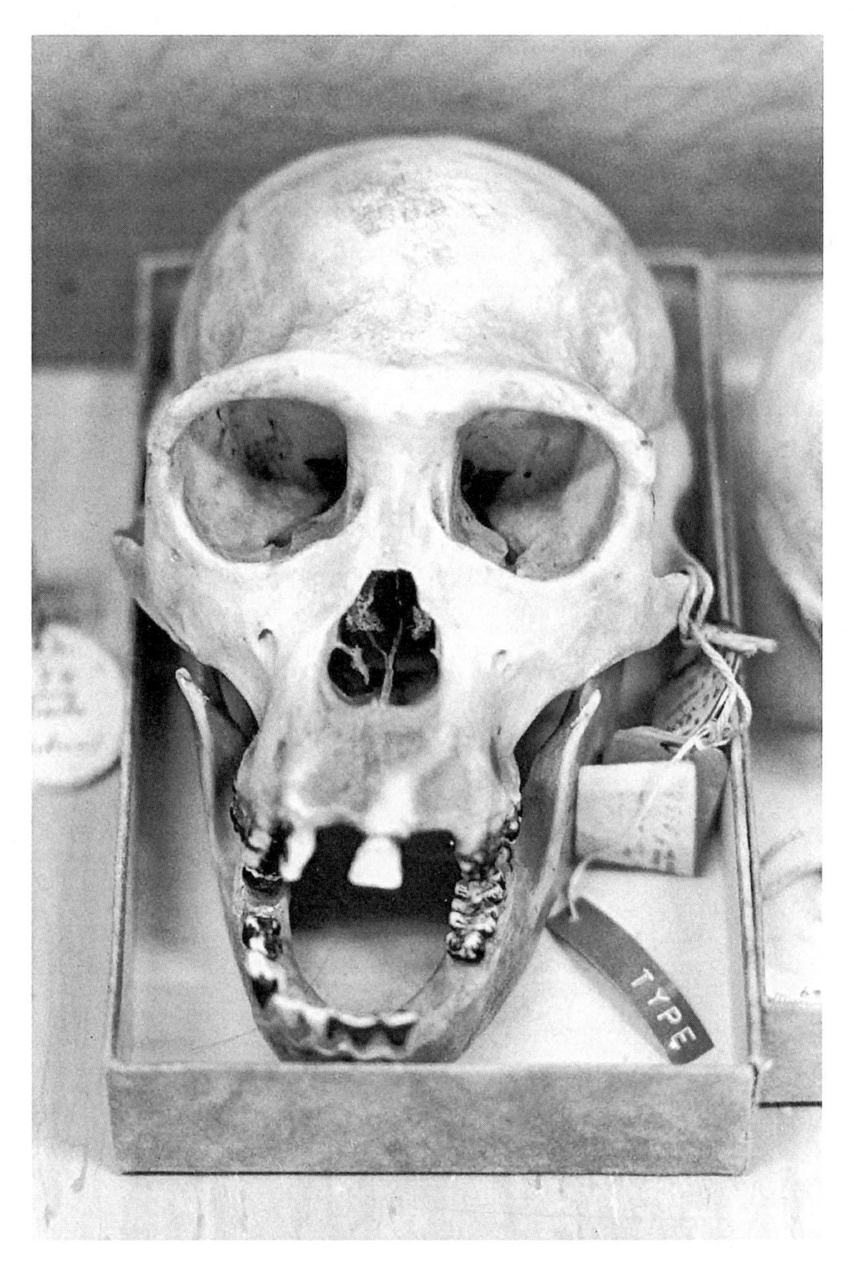

Type specimen of the bonobo (*Pan satyrus paniscus*), Royal Museum of Central Africa, Tervuren, Belgium.
(Photograph © Chris Herzfeld)

TWO

Skeletons, Skins, and Skulls

Apes in the Age of Colonial Expansion and Natural History Collections

IN THE NINETEENTH CENTURY, THE colossal task of assembling and ordering all the objects of the world was at its peak. Specimens flowed in from every corner of the European empires. However, the number of orangutans and chimpanzees that arrived in Europe remained limited. When they survived the voyage, they lived in private menageries or zoos. When they were killed for their skin and skeleton, or when they died during their long odyssey, they were put in museums to complete the collections of great apes specimens, which were as yet not very extensive. Natural history, dominated by professors positioned at the heads of museums, made little progress in the study of primates in their natural habitat. They were seen as objects of science only through the prism of their bones and remains. Carefully measured, described, and classified, reified and controlled, primates existed only through these descriptions, measurements, dissections, anatomical plates, and illustrations.

Split up into increasingly specialized disciplines, science had become professionalized. Within the framework of museums, primates would have a strange destiny. They would soon be represented essentially by a single part of their bodies: their skulls, whose facial angle would be promoted as the taxonomic criterion par excellence. By means of osteology, apes would become involved in the question of racial hierarchy, opportunely placed at the center of scientific preoccupations, while Europe underwent an unprecedented period of colonial expansion. The skull later became as important for primate specialists as stone tools were for scholars of prehistory.

Identification, Denomination, Classification, and Hierarchization

Scholars at the beginning of the nineteenth century had inherited the obsession with classifying the objects of the world that had reigned in the eighteenth century. Museums of natural history and their collections looked upon themselves as major spaces of the production of knowledge. Naturalists scrutinized organisms and made detailed accounts of their observations. Their first task was the identification of specimens and their classification in the great chain of being. Gathering all available knowledge of the great apes, scientists then compiled, in books or essays (even though they criticized the compilations, copies, and "copies of copies" of some of their predecessors, which often had false or approximate elements), their names, morphological and anatomical descriptions, remarks on their habits and behaviors, illustrations, and anatomical engravings. Invoking their predecessors, notably Linnaeus and Buffon, their studies conferred upon the animals the status of objects of science and brought them into a system of knowledge shared by a scholarly community.

Navigating between fixism and transformism, between the "lumpers" and the "splitters" who argued over the concept of what constitutes a separate species, the classification of primates remained confused until the middle of the twentieth century. The job of naturalists was made all the more difficult since specimens of the same species could appear to be very different. Males possessed traits distinct from females. Primates had different appearances depending on their age or the region from which they came. The variations were often very pronounced from one individual to another. Scholars, aware of this variability, endeavored to form collections comprising a number of representatives of each species in order to account for all the variations. However, at the beginning of the century, they had only a very limited number of specimens. Mixing names and descriptions, they had great difficulty establishing a classification of the great apes.

The Orangutan of Friedrich von Wurmb

The famous case of "Wurmb's pongo" is a classic example of this difficulty. It was known only from its skeleton and a description written by the Dutch naturalist Friedrich von Wurmb in 1780. Étienne Geoffroy Saint-Hilaire, a young French specialist in mammals, became interested in this skeleton, which had arrived from the Netherlands and was conserved at the Muséum National d'Histoire Naturelle in Paris. It was the first primate as tall as a human (measuring a little less than fifty-one inches—that is to say, the size of some individuals in certain populations such as the Pygmy peoples) that European scholars had the opportunity to study.

Geoffroy Saint-Hilaire described this exceptional specimen in "Histoire naturelle des orang-outangs" (Natural History of Orangutans), published with his friend and colleague Georges Cuvier in 1795, shortly before his departure for Egypt as a member of the scientific commission of Napoléon Bonaparte's expedition. Although the book was on orangutans, neither of the authors had ever observed a living specimen. Inspired by Linnaeus, they saw their work as trying to establish a "grand catalogue of living beings" divided into genera, order, and class, relying on precise methods and rigorous reasoning.[1] In this type of inventory, scholars do not simply name the species but also classify them in relation to their proximity to humans within the logic of the *scala naturae,* or great chain of being.

The two scholars placed the *orang-outan* of Camper and the *jocko* of Buffon just below human beings. However, they relegated the Wurmb's pongo next to mandrills among the baboons, on the next-to-last rung.[2] How to account for this classification on the lowest rung of the scale when the primate was closest to humans in form and size? Cuvier and Geoffroy Saint-Hilaire explained that the criterion to which they gave the most weight was the facial angle developed by Camper. This angle allowed, in effect, the measurement of the degree of prognathism, or the degree of elongation of the face. The sharper the angle, the more the face is lengthened forward. The further the face is from that of humans by this muzzle-like elongation, the closer it is to the animal, the brute. The demonstration seemed to be indisputable. Using a "simplified" facial angle in relation to Camper, who developed this method, the naturalists privileged chimpanzees and small orangutans, assigning a lower classification to Wurmb's pongo (a large orangutan). In 1791, however, Camper himself revised the classification of Wurmb's pongo, placing it just below the orangutan.

An Ape the Size of a Human

But what was this mysterious ape that obsessed the naturalists of the nineteenth century? We know today that it was simply an orangutan from Borneo. As we have already pointed out, naturalists at the time had been able to study only relatively young apes, which were thus small in size. Dying after several months in captivity, these primates never reached adulthood. The scientists, not taking into account their juvenile state, thought that this size was characteristic of the species. Not knowing what to make of the fifty-one inches of Wurmb's pongo, they classified it as the representative of a new genus. How to explain, then, that the animal was classified by Cuvier and Geoffroy Saint-Hilaire in a rank far removed from orangutans, when in fact it was an ape of the same species but the size of a human? It should be recalled, first of all, that at the beginning of the nineteenth century the perceived threat of the degradation of humans to the rank of animal was very much alive. Placing humans, apes, and monkeys together in the category of "Primates" coined by Linnaeus was actually forbidden in France, under the pretext that further study would prove it inaccurate.

Fears of a regression toward bestiality, revived with the arrival of apes in Europe, pushed scholars to limit the resemblances between humans and primates and to clearly define the criteria that demarcated the two, notably the angle of the face, the superior mental faculties, and the use of language. Cuvier and Geoffroy Saint-Hilaire thus qualified primates as "gross caricatures of the figure of man."[3] For them, the resemblances were strictly physical and organic. Camper's instruments and methods, opportunely revised, abetted this ideology. Wurmb's pongo was "scientifically" placed lower on the scale. Humans thus remained unique in their genus, superior to all other living beings. While orangutans and chimpanzees resembled humans, they were too small to be considered close. Many other scholars, including Lacépède, Latreille, Tiedemann, and Illiger, also classified this large primate near the bottom. In 1812, Geoffroy Saint-Hilaire, followed by Cuvier, demoted it further, placing it in the bottom rank. He would not change this classification until 1828.

Pongo, Jocko, and Wurmb's Pongo

During the editing of their work, Cuvier and Geoffroy Saint-Hilaire borrowed a great deal, without noting it, from the amateur naturalist and illustrator Jean-Baptiste Audebert. It is indispensable, however, to cite this author, since he left us with a major work entirely dedicated to the different species of primates.[4] The magnificent colored engravings, rigorous and exact, used to illustrate the book make *Histoire naturelle des singes et des makis* exceptional. Audebert espoused the principle of representing and discussing only organisms that he had seen "in nature." His illustrations and anatomical drawings thus constitute an "empirical basis" for his text and are essential elements that make the recognition of species easier.[5] He mentions three species of "tail-less" primates: the pongo (a young chimpanzee), the jocko (a young orangutan), and Wurmb's pongo (an adult orangutan). Contrary to Cuvier and Geoffroy Saint-Hilaire, he classified them in the first family of primates closest to humans. Relying on the work of various authors, Audebert noted that the pongos (called "chimpenzee" in Angola) made "huts that they covered with leaves," showed an attachment to the humans who took care of them, could eat with utensils, and were able to perform various chores, such as carrying water or turning a spit.[6]

Audebert also reported Vosmaer's observations of the second primate, the jocko, called "the man of the woods" by George Edwards. He was incorrect, however, that the jocko (an orangutan) was the same species as the primate dissected by Tulpius (a chimpanzee). He was also mistaken about Wurmb's pongo, which he believed to be different from the other two species of apes. We now know that he should have named it jocko and put it in the same group as the young orangutan. In 1818 Henri Ducrotay de Blainville affirmed that Wurmb's pongo was a subadult orangutan.

In 1835 Richard Owen confirmed that the specimens gathered under the names of "jocko," "Wallich's specimen," and "pongo" all belonged to the same species, that which we know today by the name of orangutan, which he clearly distinguished from chimpanzees. After having made a detailed account of the common characteristics shared by humans and apes, he classified orangutans in the second rank, below chimpanzees, thus restoring their position as close to human beings. In 1853 important discussions took place at the Académie des Sciences in Paris about the classification of apes. The problem was complicated by the number of names given for the same species as well as the inverse, the attribution of the same name to different species. This was the case for "pongo" and "jocko." In his *Histoire naturelle,* Buffon designated the chimpanzee as "jocko" and the orangutan as "pongo." In the supplements to this work, however, he reversed this attribution: the chimpanzee became "pongo" and the orangutan "jocko." Applied alternately to the primate from Africa and to that from Asia, the name "pongo" was also used by Battell to refer to gorillas in 1613. Today, it is used as the scientific name for the genus orangutan, whose two species are the Bornean orangutan (*Pongo pygmaeus*) and the Sumatran orangutan (*Pongo abelii*).

Orangutans, Chimpanzees, Gorillas, and Bonobos

The nineteenth century was thus troubled by the classification and hierarchization of primates. The table of species was constructed slowly, punctuated by uncertainty and hesitation, and even sometimes by steps backward, as with Wurmb's pongo. Gorillas, barely mentioned, were often hidden under the vague denomination of "Battell's pongo." They were not officially described as the second species of African ape until 1847 by Thomas Staughton Savage and Jeffries Wyman, who summarized the different species of apes known at the middle of the century. They named four "Anthropoid Simiae." Three came from Asia: the *Simia satyrus* (the immature orangutan), the *Simia wurmbii* (subadult orangutan), and the *Simia morio* (described by Owen). Only one species was believed to live in Africa, the *Troglodytes niger* (chimpanzee), until the description of a second type of ape, the *Troglodytes gorilla* (gorilla). At the turn of the twentieth century, other names of apes were still common: the tchégo, the koolo-kamba, the bald chimpanzee, the chimpanzee-gorilla, the pygmy gorilla, the pseudogorilla. Their classification remained uncertain.

In 1903, the German zoologist Paul Matschie published a description of the mountain gorilla, or *Gorilla beringei,* named as a new species in honor of the German officer Friedrich Robert von Beringe, who was one of the first Westerners to see them in the field. Ten years later, Daniel Giraud Elliot officially described two subspecies of orangutan, distinct but close cousins, the one from Borneo and the other from Sumatra.[7] In 1929, Harold J. Coolidge, a student of Robert M. Yerkes, a primate specialist at Harvard and member of the Harvard Medical School African Expedition (1926–27), published an article in which he undertook a revision of the classification of gorillas: he grouped the various species of gorillas previously described into a single species, which he divided into two subspecies, the western gorilla and the eastern gorilla.[8]

The same year, the fifth species of anthropoid primate, the bonobo, was described as a subspecies under the name *Pan satyrus paniscus* by the German anatomist and zoologist Ernst Schwarz. The bonobo was given the status of a distinct species in 1933 by Coolidge. At the time, only a handful of scientists and scholars knew of Coolidge's work and the existence of the new ape, although later this discovery came to be considered one of the most important events in zoogeography of the century. All species of apes were finally known and described. Today, the great apes are classified in six species: the chimpanzee (*Pan troglodytes*), the bonobo (*Pan paniscus*), the western gorilla (*Gorilla gorilla*), the eastern gorilla (*Gorilla beringei*), the Bornean orangutan (*Pongo pygmaeus*), and the Sumatran orangutan (*Pongo abelii*).

From Ape to Human

The work on the classification of apes was inextricably tied to the debates over evolution and transformation. Geoffroy Saint-Hilaire supported the ideas of evolutionists, while Cuvier defended a fixist (nonevolutionary) vision. Jean-Baptiste Pierre Antoine de Monet, Chevalier de Lamarck, attempted to get beyond the fixed and essentialist vision of life that had been developed in the preceding centuries. He advanced the idea that species evolved and were transformed, beginning with the most imperfect and simple and ending with the most perfect and complex. Preceded by Jean-Claude Delamétherie, he defended the idea of a simian origin for humans. Lamarck stated that men and "orangs" (a genus that comprised the "apes of Angola"—in fact, chimpanzees—and the "eastern orangs"—orangutans proper) had a common origin. These species had gradually acquired specific characteristics through transmutation, principally, as concerned humans, the ability to walk upright on two feet (bipedalism) and language.[9] Affected by the debates over Wurmb's pongo, he chose the "orang of Angola" as ancestor, claiming it to be the most "perfected" and closest to humans (including in intellectual capabilities), rather then the "eastern orang." Originally quadrupeds, chimpanzees would gradually become bipeds. Being themselves part of the animal kingdom, humans were placed at the head of the group of primates. Lamarck thus directed biology toward a systematic examination of variations among living beings within a materialist and evolutionary perspective.

The Lamarckian proposition linking the origins of humans to chimpanzees was, however, unacceptable, and even unthinkable, at the time because of the pervasive worldview that humans had been made in the image of God. Nevertheless, there were numerous examples that showed the intelligence of apes as well as their extraordinary ability to adapt to human worlds and even to take on some of their trappings. Lamarck based his remarks mainly on those of Virey, Latreille, and Grandpré, who wrote about a female ape that had learned to take care of the stove of the ship she was traveling on. She controlled the temperature and even watched out for hot coals escaping, warning the cook if danger seemed imminent.[10] Among the few naturalists to have tested the cognitive capacities of apes and monkeys, Frédéric Cuvier said that their intelligence was superior to that of a child of the same age. None of this mattered: it was thought that the intellectual and physical limits of these primates made any proximity to humans absolutely grotesque. A number of naturalists also stated that the ability for reasoning and the sociable character of the young primates gave way to a certain brutality as they got older.

In 1859, the publication of Darwin's *Origin of Species* gave rise to a renewed interest in the great apes. Between 1865 and 1870, the question of evolution, the simian origins of humans, and our place within nature excited very lively discussion at the Anthropology Society of Paris.[11] Led by Blainville, the spiritualists refused the idea that humans had emerged from a transformation, however gradual, from apes, and could not even accept that humans be reduced to their animal dimension: "Everything in an ape's form comes from some kind of material accommodation to the world, as opposed to the human form, which shows a superior accommodation for intelligence."[12] The role of apes was thus to remain in nature, and even in the trees, while that of humans was to assume a special mission in the world. Their reason and social life justified the creation of a separate "human kingdom." Different natures, different destinies.

"Man's Place in Nature"

In the face of this opposition, anthropologists claimed that studies of humans had been fettered by theology. They adhered to the Linnaean vision that made humans an object of zoological study, placing them among the animals and in the same family as apes. The biologist and comparative anatomist Thomas Henry Huxley brought together all that was known of "man-like apes" in an essential work: *Evidence as to Man's Place in Nature,* which appeared in 1863. He related the history of their discovery, described their anatomy in detail, discussed their behavior, and did comparative research on different primates, among which he classed human beings. He showed that the distance between so-called inferior monkeys and superior apes was greater than that between the latter and human beings.

In 1869 the anthropologist and naturalist Paul Broca ended the controversy with a work "of an unequaled importance to this date in scientific literature."[13] Broca affirmed that the religious bases that placed humans above all other things, isolated and central atop the great chain of being, were outside the realm of science. Holding thus to the naturalist and classificatory arguments, he confirmed in great part Huxley's conclusions and placed humans in the order of Primates. The Linnaean order of Primates was definitively restored in 1870. Nevertheless, if by the end of the century some accepted that humans were "primates," they did not accept that they were apes, even if "anthropoid."

Fallen Angels, Degraded Humans

Rejecting the transformism of Lamarck (and thus the transmutation of species), Charles Darwin nevertheless shared the other's ideas on the evolution of humans. It was not until the publication of *The Descent of Man, and Selection in Relation to Sex* in 1871, however, that the position traditionally attributed to humans among living beings would be called into question. Darwin explained the continuity between humans and apes by the probability that they shared similar ancestral forms. Moreover, he supported the idea that their differences were only of degree, not of nature. Maintaining that there were a number of common characteristics between humans and animals, he went even further in *The Expression of the Emotions in Man and Animals* (1872). There, he explained that the mental continuity between humans and apes could also be seen in their faces and bodies. This proximity was all the more threatening since all of Darwin's books were enormously successful among both an informed audience and a lay middle-class readership. Apes became a veritable "cultural obsession."[14]

With Darwin, the spotlight was focused on primates, which were once again involved in the question of human origins, a proposition that could not fail to create a scandal and offend religious sensibilities. Human beings were in effect threatened with losing their status as superior beings created in God's image, relegated to the rank of animals. The creationists denied the concept of evolution, replacing it with that of "devolution." In their eyes, if humans were fallen angels, apes were degraded men, and both were caused by an act of disobedience to God. Symbolizing the risk of degeneration, apes incarnated sin and the divine punishment inflicted upon overweening human pride. In support of Darwin, Huxley defended the idea of continuity and a common origin. Charles Lyell, a reader of Lamarck, introduced the idea of a "missing link" within the deist logic of a great chain of being. His work on geology and the age of the human species greatly influenced Darwin. The great British anatomist and paleontologist Richard Owen, however, opposed Darwin's theories. He believed there was an insuperable gap between apes and humans. For Owen, the essential difference lay at the level of the brain. He thus continued to defend the unique dimensions of humans, confining evolution to other species.

The first remains of a Neanderthal man were discovered in 1829 in what is now Belgium. Other remains were found in Gibraltar in 1848. However, it was not until 1856 that Herman Schaaffhausen described them as belonging to a separate human species. Schaaffhausen officially designated the osteological remains found in the Neander Valley in Germany as the type specimen under the name of *Homo neanderthalensis.* There was as yet, however, very little fossil evidence to support Darwin's theses. A student of Ernst Haeckel (author of the "recapitulation theory" and more interested in embryology than in the search for the "missing link"), the Dutch paleontologist and geologist Eugène Dubois began the search for a transitional fossil between ape and human. Traveling with the Dutch Army to Indonesia in 1887, he discovered "Java Man," or *Pithecanthropus erectus.* He presented Java Man as an intermediary creature between apes and modern humans, thus confirming Darwin's hypothesis of a common origin. He declared that he had finally found the famous "missing link," the transitional fossil between humans and apes.

"Let Us Hope It Is Not True"

This proven parentage filled the more conservative and religious part of the population with terror and disgust. "Let us hope it is not true, or if it is, let us hush it up," was one reaction.[15] Primates had retained a satanic image, which Christianity had embellished, up until the beginning of the twentieth century. Representations continued to show an animal of proverbial ugliness, a hideous and lascivious creature, and a figure of bestiality that was particularly detested during the Victorian period. The public was offended by the "indecent" behaviors of primates in zoos, particularly their exhibition of their genital organs and their open copulation. With profound aversion, they thus rejected any idea of a family resemblance with such beings. The physical and anatomical similarities between humans and apes troubled their consciousness and brought about debates, but most people hoped that the "divine majesty" of humans would be recognized, inviolable and triumphant.

Within this framework, the world was divided into two separate and opposing poles: culture and nature, humans and animals, civilized and bestial, with this last category including certain people judged primitive and apelike. Repulsion toward the creatures Darwin had presented as species sharing a common ancestry with humans was reflected in different works of art, for example, *The Orang-Outan Strangling a Savage of Borneo* (1895), a sculpture in the naturalist style made by Emmanuel Frémiet and exhibited at the Muséum National d'Histoire Naturelle in Paris. Like other examples of the artist's works, the sculpture tries to take into account the scientific knowledge of the time but also reflects the stereotype that the Asian ape incarnates, when adult, ferocity and danger.

The Rediscovery of the Gorilla

The fear aroused by the hypothesis of the simian origins of humans was reinforced by the arrival on the public scene of a forgotten primate, presented as being unimaginably ferocious—the gorilla. Thomas Edward Bowdich was sent by the British government to negotiate with the Ashanti, in what is now Ghana. He availed himself of the opportunity to visit several other countries in Africa. In 1817, he spent some time in Gabon. Upon his return, he described several different species of primates, including the *ingena,* most likely a gorilla, which he differentiated from the *inchego,* or chimpanzee.[16] His account, however, was ignored by zoologists, undoubtedly because it appeared in a travelogue. Regularly navigating the Congo River in the 1820s, Captain George Maxwell also had the opportunity to observe an ape whose size and strength surpassed that of a chimpanzee. In 1836, Captain Thouret brought back a skin from Africa and gave it to the museum at Le Havre in France. He was among the first Europeans to see gorillas in their natural environment, along with Paul Belloni Du Chaillu, who had the opportunity to observe them several times toward the end of the 1850s.

Encouraged by his editors, Du Chaillu, the French explorer (and naturalized American) changed his first, fairly realistic text into stories designed to impress the general public. He thus described the gorilla as the *man of the forest,* half devil and half human, hairy and savage, with powerful teeth and fiery eyes, emitting terrifying roars. His picturesque and sensationalist description, although denigrated by biologists, was all the more influential since it would remain the only one for half a century. The image that he drew was falsified, in addition, by the fact that his observations were made during hunting expeditions. Faced with the threat of a hunter, adult male gorillas protect their family. These defensive behaviors, while certainly very impressive, were interpreted as aggressive. Du Chaillu nonetheless gave some correct information. He came to know the species not only as prey but also as something like a pet, since he several times collected young gorillas when he was in Africa.

The Birth of a Giant

In 1846, several years before Du Chaillu's observations, the American doctor and missionary John Leighton Wilson discovered, in the region of the southern basin of the Gabon River, the skull of an ape that seemed to be an object of fear among the local populations. They told him that the skull was that of a primate they called *ingena.* The size, the bony crest, the eye sockets, and the impressive canines of the specimen made a strong impression upon Wilson, who presented it as belonging to a frightening beast—without ever having seen one. Wilson's host, the American missionary Thomas Staughton Savage, a doctor at the Yale Medical School and member of the Boston Natural History Society, examined the skull. A frequent traveler to Africa, Savage had published an article about chimpanzees and their habits in 1842. He was thus in a position to tell the difference between the skull of a chimpanzee and that of a gorilla. Inquiring among the local populations, he obtained other bones and sent some of the specimens to Richard Owen.

Back in the United States, he collaborated with the American doctor, anatomist, and naturalist Jeffries Wyman, who would help raise the Department of Anatomy at Harvard to the rank of a world-class institution, and who later became the first curator of the Peabody Museum. In 1847 Savage and Wyman published an article reporting the extraordinary discovery of a new species of ape. The gorilla, although it had been written about circa 500 BCE by Hannon, had been forgotten for centuries, and it was now finally officially described under the scientific name of *Troglodytes gorilla.* Chimpanzees at the time were named *Troglodytes niger.* Gorillas were thus thought of as belonging to the same genus as chimpanzees, albeit in a much more redoubtable version. One year later, Owen gave in London the first complete account of the new species, which he proposed naming *Troglodytes savagei* in honor of Savage. Owen would later conduct a number of studies on the morphology, natural history, and behavior of gorillas. He wrote an important essay about them in 1865 that for the first time included a table comparing the measurements of the four so-called anthropoid primates then known: orangutans, chimpanzees, gorillas, and human beings.

Western scientists also began field collections in order to have a sufficient number of specimens. The son of Étienne Geoffroy Saint-Hilaire, Isidore Geoffroy Saint-Hilaire, classified the gorilla in a new genus named *Gorilla,* composed of a single species and distinct from chimpanzees. Having initially named it *Gorilla savagei,* he later gave it the name *Gorilla gina* (the second part of the name coming from the local name *gina* or *engina*). He gathered all available data on gorillas in an important monograph. In 1855, a living gorilla arrived for the first time in England. Jenny was, however, represented as a large chimpanzee. She left for the United States in 1858, the year before Darwin published *Origin of Species.* Twenty years later, a young gorilla was exhibited at the Crystal Palace. It was presented to visitors by Alfred R. Wallace, who distinguished it clearly from chimpanzees, also from Africa, as well as orangutans, which he had observed in Asia. In 1884, Henri Milne-Edwards described the character of the first gorilla kept at the Ménagerie of the Jardin des Plantes in Paris and compared it to the other species of primates, which were said to be sweet and sociable, whereas the gorilla was, according to him, "savage, morose, and brutal."[17]

By the end of the century gorillas had been dissected and described by various naturalists (Hartmann, Blainville, Geoffroy Saint-Hilaire, and Duvernoy, among others). At the beginning of the twentieth century, Robert Mearns Yerkes was among the first (along with Akeley) to finally counter the gorilla's reputation for ferocity. After conducting the first systematic study of a gorilla, a female named Miss Congo in the United States from 1926 to 1928, he refuted all the fantastic stories of the past.

King Kong, Gargantua the Great, and Several Other Figures of the Savage

Gorillas continued, however, to incarnate bestiality and savagery, characteristics reinforced by their large size (adult male silverbacks can be nearly six feet tall) and massive stature. A legend heretofore associated with orangutans was also attributed to gorillas. The orangutan's reputation as a sexual predator, reported in different travelers' tales (and later corroborated in some cases), was mistakenly projected onto the gorilla. At a time when the first feminist movements were beginning to arise, women were depicted as prey who could not escape bestial violence. Once again, it was the criterion of weakness that was invoked to keep women in their "proper place." The gorilla's bad reputation was conveyed by popular imagery and took hold in the collective imagination. An 1887 sculpture by Frémiet, *Gorilla Carrying off a Woman,* speaks to the terror manifested by people faced with a potential common origin with this savage, licentious, dangerous rapist—all traits that threatened humans also and made even more glorious humanity's victory over the brute.

In the context of the Great Depression, the film *King Kong* in 1933 confirmed this powerful image while also complicating it with an ambiguous dimension: the giant was terrifying, but he was also capable of sentimental love, fidelity, and loyalty. This representation of the primate, more nuanced, undoubtedly echoed the first observations of gorillas undertaken in the field, particularly by Carl Akeley. Beginning in 1923, the great naturalist and taxidermist raised the veil on the behavior of gorillas, revealing their devotion to family. In *King Kong,* however, women continued to appear as victims and sexual objects (even though the Nineteenth Amendment in 1920 had finally given them the right to vote in the United States). The death of King Kong represented the triumph of civilization, rationality, and culture over the savage world, bestiality, and nature.

Starting in 1937, another figure, made monstrous to attract crowds, fed the public imagination. The owners of the famous Ringling Bros. and Barnum & Bailey Circus turned the peaceful gorilla Buddy into the most ferocious creature ever captured by humans. Renamed Gargantua the Great, he appeared even more terrifying due to a scar that deformed his face. Although known today mostly through very impressive posters, the reality of this representation is beyond doubt.

This adverse reputation continued until the 1970s, when Dian Fossey showed gorillas to be peaceful, calm, and attached to their families. The new image of the species as eminently peaceful was not necessarily in accord with that of the local people who lived near the primate in its natural habitat and who sometimes found themselves face-to-face with these creatures that frightened them while they were moving with their families. Gorillas nevertheless have gained a more positive image within their native countries, thanks not only to educational campaigns that have allowed people to understand them better, but also because of the enormous benefits generated by ecotourism, particularly those tied to the mountain gorillas.

The Last Ape

It was not until the start of the twentieth century that the fourth species of great ape was described. The bonobo began its career in the collections of the Belgian Musée du Congo (today the Musée Royal de l'Afrique Centrale de Tervuren), within the particular culture of natural history museums. The first encounter between this species and scientists thus did not take place in the luxurious forests of Africa but in a comfortable study in a residential neighborhood of Brussels.

It was during an examination of five skulls thought to belong to chimpanzees that Ernst Schwarz, a well-known German anatomist and mammalogist, discovered a fundamental difference compared with the other skulls in the museum: these skulls were significantly smaller than the others even though they belonged to adults, as evidenced by the fused cranial sutures. The skulls became points of reference, retaining the salient traits of the species that the taxonomist could at leisure examine, manipulate, measure, compare, and classify. It was from the evidence of these skulls that Schwarz published in 1929 the official description of this new ape as a subspecies of chimpanzee under the denomination *Pan satyrus paniscus.* One of the skulls was selected as the type specimen—that is, the unique reference in the world—that defines the "pygmy chimpanzee." The bonobos are themselves only the representatives of this type. Each species officially described thus goes back to a type specimen (which can be a skull, skeleton, bones, or a taxidermied animal), a stable element conserved in a safe place, where all members of the scientific community can go to consult it.

Skull-Being

The type specimen, associated with a denomination, a place within the classification, and an official description, becomes an archetype. In this way, only an infinitesimal part of the field enters the laboratory of the museum's Department of Zoology: the skull is charged with encapsulating an entire being and representing an entire species. In the nineteenth century, this capital element of the skeleton took its place as a major object in the construction of knowledge about primates. The absence of field studies, the rarity of great apes in captivity, and the hegemony of taxonomic work led biologists to study remains and skulls rather than living animals. This orientation of the life sciences allowed, moreover, the maintenance of a radical separation between the scientist and the object of study, supposed to assure a degree of objectivity. It made all interaction between them impossible. Craniology became in a way the supreme discipline of the time.

But why the skull? A solid, long-lasting element, it was first of all easier to transport from distant colonies than an entire skeleton, which was also difficult to reassemble once it arrived at its destination. Skulls also contain a lot of information, being the seat of four of the five senses (sight, smell, hearing, and taste). Furthermore, its symbolic status, implicitly attached to the idea of "mind," weighed heavily. The jaw and teeth also constitute important elements, as well as being an "animal" part of the human skull, since they are linked to predation. Finally, the level of intelligence was supposed to be in direct correlation to the volume of the cranial cavity. Throughout the nineteenth century, craniology (the study of the shape and size of the cranium) was immensely popular. Its assumptions would moreover serve as a basis for "scientific" theories affirming racial inequality, with the support of the facial angle of Camper, already used for the classification of primates.

The Facial Angle Theory of Petrus Camper

In the treatise *The Orang-Outang and Several Other Simian Species* (1779), Petrus Camper described the degree of prognathism (or protrusion of the lower jaw) of primates. He refined his theory of the facial angle without attaching to it any ideological determination. An antislavery advocate and defender of the uniqueness of the human species, Camper in no way wanted to provide support for the idea of racial segregation in the air at the time. Developed in 1768, this theory would not be published until after his death in 1791. Camper tried, essentially, to define the European aesthetic idea of the "Beautiful" and to place it in a broader perspective. He affirmed that the appreciation of beauty is a cultural rather than a natural construct: criteria for the definition of beauty exist, but they differ from one sociocultural group to another.

Although Camper's propositions arose from his refusal to classify, they would give way to a hierarchical scale of being that overturned the classification of primates and legitimated a taxonomy of human ethnicity based on a sole criterion: facial angle. Those who used Camper's propositions to support discrimination made "beauty" into an indicator of superiority. The ideal human was made manifest in ancient Greek statuary. Representing aesthetic perfection and thus the norm, the faces of these ancient works of art, whose facial angle is one hundred degrees, constituted the points of

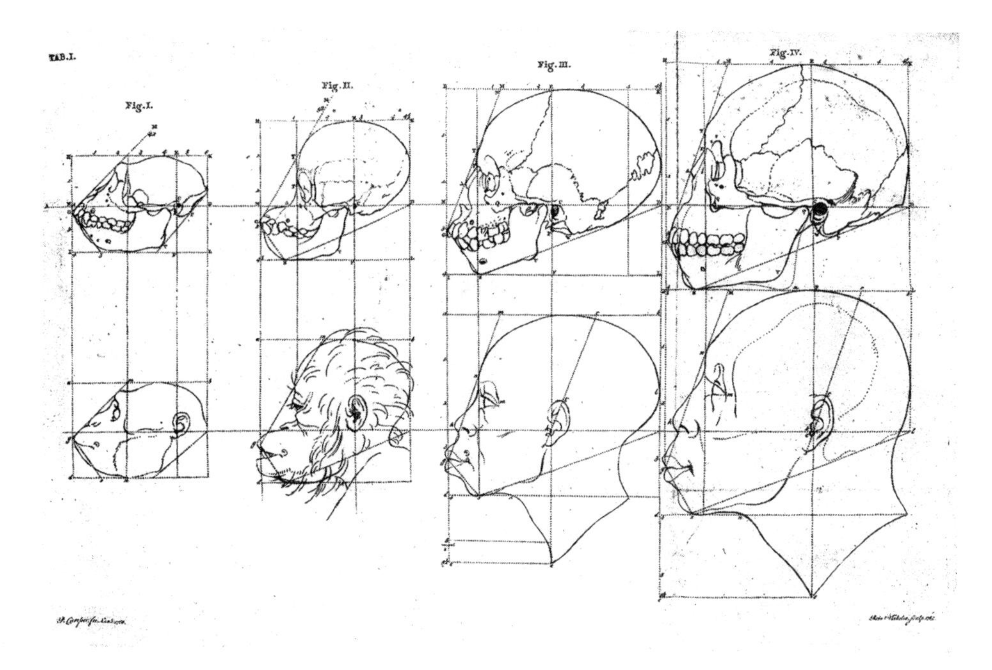

Petrus Camper's comparison of primates' skulls based on facial angle. (From Adrien Camper, *Dissertation physique,* 1791)

reference by which to measure a variance, ranging from the European (a seventy- to eighty-degree angle) to simians (a fifty-eight-degree angle for orangutans).

A Hierarchy of Races

Presenting itself as based upon an indisputable mathematical law (the measure of angles), the proportional scale established from Camper's theory was exploited by racist science in the nineteenth century: the size of the brain, intelligence, sensitivity, brutality—all could be ascertained from the amplitude of the facial angle. The great apes were brought into these studies that sought to legitimize the racial hierarchy established by the great expansionist nations. The new method of calculating the facial angle developed by Cuvier with Geoffroy Saint-Hilaire in 1795 allowed the further shrinking of the distance between certain primates and various populations with pronounced prognathism. Using this simplified angle, there were fewer differences between an orangutan and an African than between the latter and a European. Certain people even went so far as to "demonstrate" that the skull of the primate was closer to the skull of a European than was that of an African.[18]

Some colonized (or on the way to being colonized) peoples thus found themselves (opportunely for the Europeans) at the bottom of the scale, below apes! Based on a morphological and anatomical study of Sarah (officially Saartjie) Baartman, better known as the "Hottentot Venus," and transformed into a pure object of science after having been exhibited as a sideshow attraction, Cuvier and Geoffroy Saint-Hilaire had already "refuted," in 1817 and again in 1824, a black origin for the Egyptian civilization, stating that the form of Baartman's skull locked black people into an insurmountable inferiority. They evoked the "pithecomorphism," or resemblance to apes, of the "black race," and said moreover that the skulls of Egyptian mummies were similar to those of Europeans. They used the same formula of "protruding muzzle" to describe the lower faces of certain populations said to be "primitive."

The geometric rationality and apparent precision of Camper's methods were reinforced by the claims of polygenist anthropology (an anthropological system that assumes the existence of many human genera) and by the authority of Cuvier and Geoffroy Saint-Hilaire. Postulating a correlation between the projection of the cranium and the degree of intelligence and sensitivity, the naturalists imposed the facial angle as the critical distinction between the human "races," with supporting comparative tables. Certain scholars further asserted that the cerebral development of some species of apes and those of so-called primitive populations were similar. Moreover, they argued for a supposed resemblance between the physical appearance of apes and indigenous peoples. The geographic proximity "logically" led to a typological proximity, later manifested in photographs showing Africans and dead gorillas side by side with the same pose in a shared environment. This displayed their similarities and linked them metaphorically, feeding and illustrating the racist prejudices of the European colonizers. Photographs of this kind appeared at the beginning of the twentieth century and were particularly popular during the 1920s and 1930s. Widely circulated, these images became ideologically useful during colonial expansion and had a lasting impact on the structure of the imperialist imagination up until the Second World War.

Affirming the inferiority of certain human "races," the German ethnographer Robert Hartmann declared, "Among some of the lower races it is impossible not to recognize a purely external and physical approximation to the simian type."[19] In a movement of zoological comparison from humans to "animals," the "higher apes" were thus brought closer to the "lower humans": Negroes, Hottentots, Aborigines, Eskimos, Tasmanians. Small-sized peoples, such as Bushmen and Pygmies, were later evoked in comparisons with common and pygmy chimpanzees. All were said to be closer to animals than spiritually elevated "humans." It was thus "entirely natural" that Ilya Ivanovich Ivanov would attempt in the 1920s to artificially inseminate a female chimpanzee using the sperm of local African men. During the nineteenth century, much anthropological discourse thus came to be confused with the field of racist ideology, with apes being used in the radicalization of a racial landscape favorable to Western colonizers.

Great Apes and Racial Stereotypes

A radical discontinuity between peoples could thus be promulgated. The mixing of races was thought of as a disgrace and a transgression of the natural order of things. People talked about the "monstrosity" of beings engendered from the procreation of blacks and whites. There was fear of a return to a primitive, animal state, a threat hanging over an offending humanity. Westerners who settled in Africa feared losing their "white" identity. Responding to their anxieties, the problem of hybridization and theories of degeneration resurfaced. They interfered with racist theories and the idea of polygenesis. The origin of blacks was thus attributed to a cross between white humans and great apes. For the biologists of the nineteenth century, the hybrid was seen as an "individual who partakes of two natures. Within this individual is thus found difference, confusion of self and other, a scandal simultaneously ontological and biological."[20]

Certain peoples thus were seen as coming from a contaminated or incomplete humanity. Beyond the appropriation of the fauna and flora of the occupied lands, Western colonizers also carried off representatives of different ethnic groups, exhibiting them at colonial or universal expositions and even in zoos, where they were shown along with wild animals. Their remains, preserved in alcohol, were also displayed in museums alongside specimens of humans with anomalous defects. Populations of far-off countries, monsters or great apes, all seemed to be part of a "subhumanity," deemed appropriate for exhibition because of a supposed inferiority.

The theory of the facial angle would not be questioned until the end of the nineteenth century, notably by Paul Topinard. Marked by imprecision and a methodological lack of rigor, tied up with ideological or career concerns, this theory could not stand up to the data gathered from the continued expansion of collections of skulls within museums. The racial hierarchy it had supported, however, remained unquestioned.

Colonial Expansion and Natural History Collections

The opportunities linked with the colonies, the curiosity maintained by the flow of specimens brought to Europe, and the extraordinary natural richness of these distant worlds allowed naturalists to pursue the Western encyclopedic project of an inventory of all the objects of the world. During the nineteenth and twentieth centuries, the obsessional project of typification of ethnic groups and classification of living things, as well as the Western capacity for amassing large collections of specimens, reservoirs of exoticism and candidates for universality, were directly tied to Western scientists' imperialist position and status as colonizers. The abilities of scholars were diverted to political and economic ends, though they found, in return, the means to expand their knowledge.

Museums and universal expositions were put to use in the goal of exhibiting Western power. They were also designed to make the colonies attractive in the eyes of the public, promote the idea of unstoppable forward progress, and show the extent of the "civilizing work" carried out on African soil. Furthermore, deciding the names of foreign species, imposing them as universal and solely legitimate, as with the classification system that organized them, gave Westerners an additional hold over colonized peoples, a material as well as intellectual appropriation. Denomination became domination. It also celebrated the foundational act of Linnaeus who, like a second Adam, saved species by naming them, effacing, in a way, the "pagan" name given by local populations.

Other ways of designating and categorizing living beings were not considered scientific, which meant they could be ignored. However, in the case of the bonobo, for example, local populations within the pygmy chimpanzees' area of distribution had long distinguished between the *edja ya muindo* (bonobo), with a dark face, and the *edja ya mpembe* (chimpanzee), with a light face, differentiating them as well by their behaviors. When they are young, the former do in fact have a dark face from birth, whereas the latter are pale. The face of common chimpanzees gets darker only as the individual ages.

From Menageries to Zoos

In parallel with their lives as taxonomic objects, primates experienced other modes of existence in the West. At the end of the eighteenth century, various species of monkeys lived in private menageries at a time when exotic animals in Europe were still extremely rare and reserved for the rich and powerful. Several collections of living animals were, however, open to the elite, as well as to a discerning public, sometimes coming from the bourgeoisie. Buffon himself regularly visited the animal parks of Versailles and Chantilly. He also established an animal reserve in his hometown, Montbard, where he stayed regularly. People had access to exotic animals, including monkeys, trained or not, through animal showmen up until the end of the century (when they were banned in France), or at traveling zoos until the beginning of the twentieth century.

Created in 1793 by Étienne Geoffroy Saint-Hilaire, the Ménagerie at the Jardin des Plantes in Paris was the first public zoo. The expense of maintaining it was justified by the pedagogical dimension of the institution, its participation in the progress of science, and its interest for naturalists and artists. The residents of the zoo became in a way a living illustration of taxonomy, the collections of living animals constituting the extension of the natural history collections made up of dead specimens. Through these living collections zoologists were able to establish precise descriptions of exotic species on a morphological as well as a behavioral level. One of the first animals to arrive at the menagerie was a primate. It was confiscated by the authorities, along with a polar bear, a leopard, and a civet, in order to assure public order and guarantee the safety of citizens.

The Orangutan of Empress Joséphine

It was at the Château de Malmaison, the residence acquired in 1799 by the wife of Emperor Napoléon I, Joséphine de Beauharnais, that Frédéric Cuvier made one of the first scientific studies of a living orangutan. Cuvier drew a detailed portrait of the animal, the first of its species to have arrived alive in France, in March 1808, including it in an 1810 publication, *A Description of an Orangutan, and Observations of Its Intellectual Faculties.* Having a certain taste for natural history, the empress established her menagerie because she wanted to acquire exotic animals in order to try to tame them. General Charles Decaen, governor of Pondicherry (India), offered her a young female orangutan named Rose. Mademoiselle Avrillion, first chambermaid of the Empress Joséphine, reported in her memoirs that a highly cultured primate had been seen by Napoléon, himself an amateur scholar of curiosities from far-off lands.[21]

Rose wore a frock coat, ate with utensils, loved turnips, and walked on two legs. She had unfortunately lost several fingers to frostbite while crossing the Pyrenees during her travel to Paris. Cuvier estimated her age to be between ten to fifteen months. He described the orangutan's physical aspect, diet, vocalizations, behavior, locomotion, arboreal character, and sentiments (affection, anger, sadness). He stated that the manner in which Rose expressed some of these sentiments could not be observed in any other animal except for humans. The scholar wondered whether these manifestations were motivated by the same causes as among human beings. He believed so, since the primate, for example, would pause in her cries of anger to judge their effect, and recommence crying out if she did not get what she wanted. The ape thus showed a real ability to evaluate the mental states of others.

Cuvier also explained that she used her hands in the same way as humans and that she had some of their intellectual abilities (judgment, an ability to generalize and to anticipate), confirming an extreme closeness between the two species. Moreover, the orangutan showed an interest in other species, becoming friends with two small cats. Finally, Cuvier noted Rose's essential need for affection as well as a desire to suckle, for which reason they tried to have Rose nursed by a wet nurse, then by a goat. The young primate refused them both. She died after five months in captivity.

Observing Living Animals

At this time, the life expectancy of captive primates was often little more than several months. After death, they were dissected and many parts of their bodies were kept in collections. The study of animal behavior was secondary to taxonomic research. Nevertheless, some researchers believed that the observation of living specimens was a necessary condition for genuinely understanding animals. Closely tied to the culture of museums of natural history, the English naturalist Edward Donovan became interested in the behavior of an orangutan from Borneo that arrived, alive, in England in 1817. The ape was kept for two years at the menagerie at the Exeter Exchange in the Strand. He drank tea, ate with a spoon, liked brandy, and showed a lot of affection for his keeper. He easily recognized the person who had brought him from Java, the doctor Abel Clark, when he came to visit, searching the doctor's pockets for sweets as he had when they traveled together. In a book complete with many illustrations, Donovan wrote nearly one hundred pages about the great apes.[22]

In Paris, Georges Cuvier's younger brother, Frédéric, named superintendent of the Ménagerie at the Jardin des Plantes, was the first to develop at the menagerie systematic research on the behavior of mammals, on "the intelligence and instinct of animals," and more specifically on the cognitive abilities of primates. From the first experiments, the primates understood what was asked of them, then repeated what they had been taught. Cuvier and Geoffroy Saint-Hilaire thus showed "that the apes appropriate, in a way, these actions that are not natural to them, and that they execute them" and "whatever the circumstances . . . once the apes have learned to drink from a glass at the table, whenever they are thirsty in between meals, they look for a glass, fill it, and drink; from then on, they only drink like men."[23]

The Ménagerie received its first orangutan on May 15, 1836. Less than a year old, Jack stayed with his keeper's family: he ate at the table, played with the children, and shared a bed with the son. Jack was observed by Étienne Geoffroy Saint-Hilaire, who published his observations as "Studies of an Orangutan at the Ménagerie." A chimpanzee arrived October 18, 1837, the year Frédéric Cuvier finally managed to create a chair for the study of animal behavior and psychology. It was also at this time that a new monkey house was built. When Frédéric Cuvier died suddenly in July 1838, the animal park passed to the authority of Geoffroy Saint-Hilaire, who was himself superseded by his son Isidore in 1841. The zoologist Henri de Blainville, having taken care of a number of primates at the Ménagerie himself, including Jack, related how attached one chimpanzee became to the people who looked after it. The young primate loved to play, jump, and swing. He was calm and obedient: "a word sufficed to make him stop, come over, and be embraced like a child; he had this in common with human young, that he did not like to be left alone, and cried when neglected."[24]

The physiologist Pierre Flourens continued Frédéric Cuvier's observations and made his own commentary about the third orangutan to arrive in France: "We were going to retire, when he again approached his new visitor, and took, gently and playfully, the cane he had in his hand, and pretending to lean on it he curved his back, slowed his pace, and walked around the room we were in, imitating the posture and walk of my old friend. He then brought back the cane on his own, and we left him, convinced that he too knew how to observe."[25] Representations of primates using a stick to walk upright on two legs may therefore sometimes have corresponded to real situations, the primates imitating their elderly human companions. Furthermore, Flourens noted that the orangutan did everything that Buffon had said about the chimpanzee he had observed. He also related an experiment worthy of the future work of Wolfgang Köhler. It was usual to lock the door to the room where another orangutan lived, the key being then placed on the chimney where the ape could not reach. The orangutan, however, got the key by using a rope attached to the ceiling. Keepers then shortened the rope by tying a series of knots. The ape did not hesitate: he untied the knots and grabbed the key. Apart from these few tentative observations of animals in zoos, few studies of animal behavior were done in France. Rather, veterinary research was privileged, mostly because of the high rate of mortality among captive animals.

Zoos

Throughout the nineteenth century, a number of European cities constructed zoos, emblems of power and new places of urban sociability. The London Zoo was created in 1828 in Regent's Park, offering a significant avenue for scientific research. The zoo received its first chimpanzee, Tommy, in 1835. Eighteen months old upon his arrival, Tommy died after six months at the zoo. In November 1837, Lady Jane (renamed Jenny), a three- or four-year-old female orangutan from Borneo, was admitted. She died in May 1839. Two other orangutans, also named Tommy and Jenny, succeeded her. Wearing a dress, the young female drank tea with a spoon with such delicacy that Queen Victoria and Prince Albert, who visited her in May 1842, were astounded. Jenny died in October 1843. Charles Darwin had been a fellow and corresponding member of the Zoological Society since 1831, the year in which he began his voyage on the *Beagle.* He encountered an orangutan for the first time at the London Zoo in 1838. He observed the first Jenny, then Tommy and Jenny (the second), and conducted several experiments on them with the collaboration of the zoo's veterinarians and keepers. He was interested not only in the primates themselves but in what they could teach him about the evolution of the human species. From 1885 to 1891, his young colleague George Romanes conducted studies on the female chimpanzee Sally, also at the London Zoo.

Orangutans would play a major role in the search for human origins in the nineteenth century. Returning from Malaysia in 1862, Alfred Russel Wallace became an expert on this species, as he was one of the few specialists to have seen them in their natural environment in Borneo. He had collected a number of specimens of skins and skeletons of orangutans, taken measurements, and noted their behaviors: building nests, movement through the canopy, diet, and their solitary nature.[26] He even had the opportunity to spend three months with the infant of a female he had killed, calling it, humorously, "his baby." He nevertheless remained a hunter and collector, keeping his distance (physically as well as symbolically) from the primates that he had occasion to encounter. For him, they remained, above all, prey, commercial and scientific objects. He also described the apes using the derogatory term that would long be associated with them: monsters.

Founded in 1844, the Berlin Zoo received its first gorilla in 1876. The most important trader of exotic animals of the time, Carl Hagenbeck, brought specimens to various European cities. He opened a menagerie near Hamburg, Germany, in 1863. A number of zoo directors, or even keepers, brought apes they had in their care into their homes, where they became playmates for their children and ate at table. This was notably the case, in the following century, for William Temple Hornaday, the first director of the Bronx Zoo (then called the New York Zoological Park, which opened in November 1899). The American zoologist and conservationist even organized experiments to test the intelligence of a number of chimpanzees and orangutans that were kept at the zoo. In addition, he had the chance to observe orangutans and to live with one of them in Borneo in 1878. At a time when people did not travel much, zoos became almost the only places where scientists as well as the general public could see the great apes.

Apes at Work

Toward the middle of the nineteenth century, several authors, including the science journalist and utopian socialist Victor Meunier, proposed putting the great apes to work. Their attachment to humans, their capacity for imitation, and the fact that they had hands were arguments in favor of this project. Moreover, work would in effect raise them up, since it already had, according to Engels, transformed apes into humans. Meunier saw animals through the prism of work, which in turn was seen as being the supreme law of living beings. Why should the animal closest to our species, with hands, not participate in the collective effort, the immense and universal deployment of activity?

Meunier declared that the use of primates by humans could constitute an emancipatory progress and a means to fight against the exploitation of humans by humans. Whether in the city or in the country, primates could be given artisanal tasks left vacant by workers employed in factories as well as certain jobs in industry, workshops, and agriculture. The males would be employed as guards, gardeners, tailors, prospectors, and sailors. The females would serve as babysitters, nursemaids, and domestics. Slavery of the masses would finally be eradicated. In order to achieve his goal, Meunier proposed creating zoological stations in Africa in order to train these future working masses.

He also gave various examples of domestic apes, including those reported by Achille Poussielgue. The chimpanzee Antonio, twelve years old, lived with General Llorente at San Geronimo in Gabon. Dressed in white pants, scarlet vest, a bright red cap, and a towel draped over his arm, he set the table and cleaned the dishes without breaking them. Meunier maintained that he did his job much more effectively than the general's black servants. He thus implicated primates in a project of humiliation of colonized peoples. To support his point of view, he also reported that in April 1885 three Bushmen had been placed next to a chimpanzee, named Jacques, in order to compare the two species. This encounter had the result, according to him, of showing the same baseness of conduct. The author proposed reducing the alienation of the Western working class, but he had no qualms about exploiting other populations that he qualified as "primitive," as well as the great apes, to achieve his goal.

Apes in the Age of Machines and Empires

What does the primatology of the nineteenth century tell us about the way of doing science and seeing the world in the age of machines and empires? Defining themselves by their superiority over other living beings, humans reified animals in the form of specimens (and thus stable elements), studied and collected in museums of natural history. Specific to the West, this way of constructing knowledge about infinitely diverse, infinitely exuberant living beings from cadavers was also applied to primates. While naturalists observed living specimens when they could, in menageries or zoos, they worked mainly on primates reduced to the state of taxonomic objects. This reification made possible the work, practiced intensely throughout the nineteenth century, of description, denomination, and classification.

Scientists were thus distant from the animals they studied as well as from their physical, social, and cultural environment. The great apes were furthermore dragged into the question of the classification of races, critical at this time of colonial expansion. The supposed division between humans and animals was used to classify colonized people on the side of the brute, of "barbarian intelligence" and ferocity. The Europeans in this justified their takeover of indigenous lands and their "obligation" to attempt to "civilize" them. The attribution of names to species living in far-off lands constituted another form of appropriation and domination by the Western colonizer.

Furthermore, without the extraordinary influx of innumerable cultural objects as well as botanical and zoological specimens from the "colonial periphery," neither the Western natural sciences nor the departments of anthropology and ethnology of the great Western museums would ever have experienced their exceptional development during the nineteenth and twentieth centuries. The objectification of primates was not limited to museums. During this period of industrial revolution, apes were implicated in class struggle: there were hopes of putting them to work in order to liberate human workers. In popular literature, they were saddled with every imaginable role, from monster to the missing link, rapist to assassin, wild man to savior of young girls. They were kept in zoos, "prophylactic doses of nature" within cities.[27] The different places for keeping the exotic—zoos, colonial expositions, and museums—were showcases for the power of the great European nations as well as evidence of their hegemony over the world, other peoples and all living things, including their closest cousins.

The chimpanzee Grande reaching for a banana placed high above her.
(From Wolfgang Köhler, *The Mentality of Apes*, 1925)

THREE

Apes as Guinea Pigs

Primates and Experimental Research

HOW WERE GREAT APES USED IN RESEARCH from the end of the nineteenth century onward, particularly in the context of the rise of experimental biology? How was the new discipline of primatology invented? Until the beginning of the twentieth century, primates often arrived in the West sick or as cadavers. Only a handful of individuals survived the long voyage after being torn from their native land. However, as time went on and the infrastructure in the colonial territories became more efficient, the number of specimens sent increased, and the primates, taken better care of, survived in greater numbers. Researchers were thus able to conduct systematic studies of primates. The extremely limited knowledge of these species needed to be verified and expanded, as much from a biological point of view (anatomy, pathology, neurology, reproduction, development) as from a cognitive, social, and behavioral perspective.

The first scientific studies of primates' behavior (ways of learning, instinct, intelligence) were conducted in private residences or at university laboratories. Then came the first research departments specifically established for this purpose: the Prussian Anthropoid Research Center on the Canary Islands (1912), the Department of Zoöpsychology at the Darwinian Museum in Moscow (1913), and the Primate Laboratory of Yale University at New Haven, Connecticut (1925). The subtropical Primate Station of Sukhum in Georgia, USSR, founded in 1927, was a breeding station for a number of Russian biomedical research institutes. Following dogs, cats, rats, and pigeons, primates entered the laboratories. Behaviorists using these eminently social beings in experiments also wanted to adhere to the dictates of the hard sciences, and thus came to treat them as standardized experimental objects. This positioning became even more radicalized in biomedical laboratories. Thus, at the beginning of the twentieth century, the experimental method (and its attendant measurements, protocols, recording of facts, statistics, and parametric controls) extended its hegemony over primate research.

Cognition, Language, and Imitation Among Apes

At the end of the nineteenth century, studies on the theme of "animal intelligence" abounded. However, the idea of a mental life applied to animals constituted a real provocation. According to the prevailing worldview, humans were the only species endowed with intellect and reason, attributes that justified their place at the top of the great chain of being. Putting this belief to the test, naturalists and psychologists attempted to explore animal cognition, particularly that of primates. They were interested in questions of problem resolution, language, tool use, and resemblances to humans. With the emergence of modern zoos in the nineteenth century, a new space of observation was opened for those interested in animals, and another site of tension developed between scientific and common knowledge.[1]

Charles Darwin postulated that there was a mental continuum, as well as similar cognitive mechanisms, between humans and "superior mammals." Based in large part on his observations at the London Zoo, which was a sort of laboratory for him, his book *The Expression of the Emotions in Man and Animals* (1872) set off a real passion for the study of animal behavior and psychology. At the time of the arrival of the first chimpanzee, Tommy, to the London Zoo in 1835, Darwin was still voyaging aboard the *Beagle*. However, he encountered the orangutan Lady Jane (renamed Jenny) and another orangutan (also called Jenny) toward the end of the 1830s. He organized experiments at the zoo with the help of the keepers. Notably, he tested the primates' reactions in front of a mirror. At first the apes took their reflection for another orangutan and appeared frightened. Contrary to humans, they seemed incapable of recognizing themselves in a mirror. On the other hand, Darwin thought that Jenny understood human speech. Darwin believed that the apes were inept at fabricating tools: "An anthropomorphous ape, if he could take a dispassionate view of his own case, would admit that though he could form an artful plan to plunder a garden—though he could use stones for fighting or for breaking open nuts, yet that the thought of fashioning a stone into a tool was quite beyond his scope."[2] He thus was aware of the ability of apes to use stones, either as tools or to defend themselves, but he denied their ability to make tools.

First Experiments

Edward Pett Thompson, a British naturalist and creationist, reported the ingenuity of the orangutan in the Ménagerie of the Jardin des Plantes in Paris. The ape moved a stool in front of the door of its cage to get at the key. Still not able to reach it, the orangutan put another stool atop the first, and escaped from the enclosure. It was thus able to string together a series of actions to attain a specific objective, demonstrating its capacity to understand the situation and respond intelligently, finding a solution without resorting to trial and error.

In an article published in 1889, George John Romanes, a disciple and friend of Darwin, recounted the behavior of Sally, a female chimpanzee who had lived for many years at the London Zoo. Sally was able to eat with a spoon. She was said to be intelligent and mischievous. She loved to play with her keepers and seemed to have a high level of comprehension of spoken language. Romanes taught her to give one, two, or three pieces of straw. Sally was rewarded each time she gave the right number. She thus learned to "count," or rather associate a sound with a certain number of pieces of straw, up to the number five. Higher than this she was less precise. Cleverly, she often broke a piece of straw in two so that she did not have to look for another. Romanes tried to teach the names of colors to Sally, but in vain. She learned to recognize only white. On the other hand, Romanes wrote of an orangutan in the Paris Ménagerie who patiently tested the fifteen keys of his guardian's key ring until he found the one that opened the door. This individual also used iron bars as levers. Romanes insisted on apes' behavioral, cognitive, and structural proximity with humans. For him, the fact that apes did not possess a sophisticated language did not mean they lacked intelligence, whereas many linked the two faculties to justify an irreducible separation between humans and apes.

Romanes rounded out his observations with various anecdotes (whose credibility he nonetheless assured). His approach, which tended more toward naturalist inquiry than scientific experimentation, ruined his reputation in the eyes of an increasingly preeminent fringe of specialists in animal psychology and behavior. Defenders of an experimental version of their discipline, these researchers advocated the rigorous application of the scientific method, the control of groups being studied, measurable variables, and large samples of objects being tested. Romanes was notably criticized by Conwy Lloyd Morgan, a pioneer of comparative psychology and author of the famous "Morgan's Canon."[3] For Morgan, Romanes's studies were compilations, stamped with anthropomorphism and sentimentalism. In addition, the psychologist was not impressed by Sally's counting. He thought she was not capable of generalizing her knowledge of numbers: she never applied it to other objects, thus showing that she had not really understood the meaning.

Considered the first ethologist in the modern sense of the term, Morgan also undertook some observations at the London Zoo around 1890, notably of a chimpanzee he named Cousin Sarah. He brought attention to the fact that the younger apes are, the closer they were likely to be to humans and to share certain refined habits with them. Morgan declared that as primates got older, they became more savage and ferocious.

Richard Garner, Moses, and Susie

Starting at the end of the nineteenth century, researchers began to address the question of language among primates. The American naturalist Richard Lynch Garner observed different species of primates in zoos beginning in 1884. A reader of Darwin, he thought that all animals had the ability to communicate to a degree adapted to their needs and experience. As opposed to the research of the 1950s that attempted to teach sign language to apes, Garner tried to understand what they were already trying to communicate. He decided to learn the language of apes.[4] His dream? To record their vocalizations. He was thus interested in Thomas Alva Edison's wax cylinder phonograph, invented in 1877. The phonograph became in this way a technical apparatus that Garner used in the service of science during the 1880s.

Starting in 1892, Garner frequently traveled to Gabon and Congo-Brazzaville (at the time French Congo). Known as "the Monkey Man," he lived alongside a number of chimpanzees. Some of them shared his table and slept in his house. In 1893 he adopted a young chimpanzee that he named Moses. He began little by little to recognize some of Moses's vocalizations and attempted to reproduce the sounds. Garner came to believe that he established communicative exchanges with the chimpanzee. Moses was also able to distinguish forms and colors and learned to "sign" a page with a cross. After Moses died, Garner traveled to Liverpool with two chimpanzees, Aaron and Elishiba. After a further voyage to Africa, he returned to the United States with a gorilla, Dinah, and several chimpanzees, among them Susie, which he tried to raise as his own child and took along in his traveling show in 1910 and 1911. He thus made his theories popular, bringing to the public the idea that apes were able to communicate, think, and reason.

Garner continued to explore the question of language among different species of primates as well as the ability to count among capuchins. He was nevertheless strongly criticized by the scientific community, not only for his methods and his hasty, unfounded interpretations, but also because he was considered an "amateur" and guilty of anthropomorphism. Because of his publications in the popular press, his lack of any scientific training, and his presentation of himself as an animal trader, adventurer, hunter, and showman, he in effect became part of the "primate folklore" that researchers tried so hard to distance themselves from. However, very few "professionals" of the time could boast of such experience in the field and such a deep engagement with primates.

Louis Boutan and Pépée

The French biologist, photographer, and explorer Louis Boutan also criticized the work of Garner, principally because of its imprecision and amateurishness. Taking part in a scientific expedition to Indochina beginning in 1904, he received as a gift a young female white-cheeked gibbon during a trip to Laos in 1907. He named her Pépée. Upon returning to Hanoi, Boutan gave Pépée to his wife, who raised the gibbon as her own daughter. The couple had no children. Céline Boutan dressed Pépée, put her to bed, played with her, taught her to use cups and tableware, and sat her in a highchair. Boutan advocated a very close relationship between the researcher and his subject. He insisted on the importance of establishing trust with the animal and making sure it had a safe place of learning. He began various experiments at his home designed to penetrate the mystery of the innate mode of communication proper to the species. He characterized Pépée's vocalizations as a "pseudo-language" and evoked another state of communication, that of "rudimentary language." Garner supported the idea that the sounds emitted by certain primates corresponded to words in human language, which Boutan refuted: for him, these sounds correlated to emotional states rather than words.

Returning to France in 1908, Boutan finally had the opportunity to record Pépée's vocalizations via phonograph. He pursued his work in an equipped laboratory at the Faculté des Sciences in Bordeaux, where he was a professor of biology and animal physiology. During this pioneering research in experimental psychobiology, he showed that certain of Pépée's spontaneous behaviors were extremely close to those of human children between two and ten years old. With the aid of puzzle boxes (whose system of opening was of varying complexity), he tested the ability of primates to resolve a problem according to their own way of doing things. He adopted certain experimental designs used by British and American psychologists. However, he did not place his subjects within an overly restrictive schema of learning, leaving them a certain degree of freedom. He anticipated the idea of insight, although somewhat tentatively. He thus supported the development of an animal psychology based on direct observation, though distanced from that of Thorndike.

Some placed Boutan between Romanes (who accorded too much to animals in terms of intelligence) and Thorndike (who, on the other hand, accorded them too little). Boutan was also criticized by those who held a more rigorous view of science, who thought he did not adequately apply quantitative methods. According to them, he remained an amateur who deserved little credit. However, Robert M. Yerkes admired his work, believing it to be an important precursor to his own studies.

Making Use of the Living

The beginning of the twentieth century was profoundly marked not only by the industrial revolution but by an intellectual one, arising from various historical, cultural, and scientific shifts. After the Ptolemaic geocentrism of the Renaissance was put into doubt, the place of humans at the center of the universe was called into question. With Darwin, human beings were seen as but one species of animal among others; following Freud, humans were forced to accept that an essential part of themselves, the unconscious, lay beyond their control. Added to these was Albert Einstein's theory of relativity. All that had once seemed sure became uncertain. To this agonizing loss of reference points and stable meanings, this shaking of foundations, intellectuals responded with compulsively controlled research and a valorization of the experimental method, which soon spread to all the disciplines, including the human sciences, in a positivist and reductionist worldview. This was also tied in many ways to the world of industrial production as well as the industrialization of agricultural production. Steeped in this climate, the life sciences took part in a belief that pushed the demands of objectivity to its limits, as bodies were subjected to multiple procedures of measurement and transformed into controllable entities.

A new relation to living beings took place. Apes and monkeys were used as experimental materials and exposed to a standardization that applied to a large sampling of species, from fruit flies to chickens, from rats to pigeons. Certain humans also were used in research that transformed them into objects of experimentation: condemned prisoners, indigenous peoples, detainees, handicapped people, prostitutes, slaves, or patients in psychiatric institutions. Some saw this extension of scientific experimentation to humans as so indispensable that certain subjects were deemed subhuman in order to justify their use. Furthermore, in between the two world wars, the close ties between science and the military-industrial complex reinforced an orientation marked by three specific factors: pragmatism, rationalization, and systematics. Science became the solution to every problem, characterized by multiple mediations, the administration of complex systems, a single point of view, and a culture of analysis, management, and planning.

When Primates Became Experimental Material

It was in this climate of scientism, positivism, and the instrumentalization of the living that animal behavior became an object of scientific study. Experimental psychology became institutionalized in the United States between the nineteenth and twentieth centuries, whereas in Europe it was still in its infancy. Under the influence of Douglas Alexander Spalding (called "the first experimental behaviorist") and John Lubbock, the procedures used in psychology became aligned with modern methods of laboratory science. They took over research dedicated to animal intelligence and behavior. In 1900, animal psychology was officially integrated as a branch of experimental psychology at Yale and at Harvard, where a laboratory specifically equipped to house different species of animals was created. Together, Robert Yerkes and John Watson published an article proposing experimental designs.

Primates, however, did not yet figure among the animals being tested. Edward Lee Thorndike at Columbia University pioneered the application of the experimental method to studies of primate intelligence. In 1899 he conducted research on three capuchin monkeys from South America at his home in New York. Two years later, he published a monograph entitled *The Mental Life of Monkeys.* In 1911 he described primates' manner of resolving problems through trial and error in *Animal Intelligence.* He thought that this process was the only mode of learning for the species under study, this learning being reinforced by a system of conditioning under the form of "reward-punishment." Passing from a psychology based on general principles to one based on precise and systematic experimental techniques and using different methods, such as the famous "cage-tests," which Thorndike perfected, research he conducted marked the beginning of the first controlled experiments using animals. Explained by a simplistic system of reactions (stimulus-reaction), complex behaviors (as well as their underlying mechanisms) were analyzed in an elementary manner under the yoke of the crucial criterion of "objectivity." For Thorndike, the learning process was the same for all species, and animals lacked the ability to reason.

Do Horses and Primates Think?

Among the researchers who questioned Thorndike's vision was the British sociologist and liberal theoretician Leonard Trelawney Hobhouse, who was not satisfied with the unique explication of learning by association proposed by various psychologists. He believed animals learned in a more complex manner than trial and error. He set forth his views in a book, *Mind in Evolution,* which appeared in 1901, the same year as Thorndike's *Mental Life of Monkeys.* In fact, both men began to work on primates at the same time. Conducting his research at the Manchester Zoo, Hobhouse tested the use of tools with a rhesus macaque, Jimmy, and a chimpanzee, Professor (he also tested dogs, cats, elephants, and otters), and showed that they demonstrated great dexterity. Contradicting Thorndike's hypothesis, he thought that learning was not always linear and gradual, but more organic. He also questioned the efficacy of progressive reinforcement of appropriate behavior through conditioning and association.

For Hobhouse, primates also recalled past experiences and precise knowledge when connecting elements they were given in new situations. He believed that these processes attained a high degree of complexity and manifested themselves differently according to the particular species. Certain individuals were capable of giving a very rapid response to a problem without preliminary attempts; the solution emerged spontaneously. Others found a good solution through imitation. Animals thus showed "rudimentary reasoning," practical judgment, and intentional action. Hobhouse was strongly criticized by proponents of experimental science who from the outset adopted Thorndike's ideas and principles: they reproached the imprecision of Hobhouse's methods, his lack of methodological rigor, the small sample size of individuals tested, nonquantifiable results, the absence of a control group, and an overall lack of objectivity.

At the beginning of the 1900s, the unmistakable performance of animals deemed "intelligent" would, however, challenge the scientific materialism advocated by Thorndike, who had gradually established a degree of hegemony in animal psychology: Wilhelm von Osten claimed that his horse Hans knew how to count. A commission of experts (among them the scrupulous Oskar Pfungst) gathered to test Hans's mathematical abilities. After many debates, the scholars showed, in 1907, that in reality the horse was responding to certain subtle signals unconsciously transmitted by his master ("muscle-reading" and "sign hypothesis") that enabled him to tap the correct answer with his hoof. He merited in a way the description "clever," but not because he knew how to count.

The Thinking Horses of Elberfeld

After von Osten's death, the "amateur scientist" Karl Krall bought Hans and formed a small colony of animals, many of them horses, known as "The Thinking Horses of Elberfeld (and Manheim)." He educated them, along with other animals such as parrots, elephants, cats, rats, and donkeys, as if they were children who had been born mute. He attempted to teach them to reason and resolve complicated calculations while communicating with humans. Krall tried to prove their cognitive abilities in a scientific and experimental manner at his laboratory in Munich, well equipped with measuring tools and modern technical means.

However, at this crucial time, when a new comparative psychology applied to humans and animals was being developed, the voices of experimental orthodoxy in the discipline did not accord Krall the official recognition he desired. They believed that his results were falsified and the experimenter duped by lack of prudence and rigor. Krall became the perfect example of the harm that uncontrolled, or difficult to control, parameters could cause. The objects of his study (consciousness, spirit, the transmission of thought) were ridiculed. The research methods used by Krall would be expunged from the field of experimental research as preeminent examples of amateurism and lack of objectivity, both closely associated with the sin of anthropomorphism. In short, Krall's work exemplified all that must be avoided in "good science," kept from contaminating a scientific discipline in the process of establishing itself. "The Thinking Horses of Elberfeld" joined the "Clever Hans Phenomenon" as a methodological cautionary tale for researchers. Reductionist experimental psychology triumphed: psychology became experimental rather than naturalist, controlled rather than anecdotal, although the ghosts of Krall's horses would continue to haunt the study of animal intelligence.[5]

The Hour of Behaviorism Triumphant

In 1913 John Broadus Watson published "Psychology as the Behaviorist Views It," today considered as the manifesto of "behaviorism." One of his goals was to avoid the errors incarnated in the Clever Hans Phenomenon. He used primates in his research as well as rats and pigeons: "Pigeon, rat, monkey, which is which? It doesn't matter," declared B. F. Skinner, another proponent of this type of research.[6] In behaviorist logic, the learning process (the center of these researchers' preoccupations) is fundamentally identical for all species, as it is fixed in a rigid framework from which the animal cannot escape. For these psychologists, who insisted upon the importance of a continuum between species, primates were no more adept than dogs or cats in reasoning or thought.

In this perspective, which placed the experimental method, prediction, and control at the forefront, it was indispensable to neutralize individual variations and work with standardized subjects. Living beings were thus fashioned and made to conform to certain norms, tied to the culture of industrial productivity, in order to assure the robustness and stability of the results, the reliability of the method, and the experimental precision. The few individuals tested were the so-called representatives of the entire species, and the results obtained could therefore be generalized. Their life history as individuals, their preferences, their capacity for initiative, and their inventiveness were reduced to nothing.

Primates were thus forced into an existence—uniform, routine, miserable, actually life threatening—that allowed them neither a social life nor the least degree of freedom. The behaviorists accentuated the differences between humans and animals, a fundamental distinction that allowed them to treat animals cruelly and pitilessly since they were supposedly devoid of reasoning, emotions, or sensitivity. The influence of such research became widespread in the world of comparative psychology and behavioral studies. The first research using primates was done by Watson, who was interested in the question of imitation and published a paper on the subject in 1908. Watson was also the coauthor of a 1913 paper on the development of a young monkey with Karl Spencer Lashley, who would further propose a study of laterality in rhesus monkeys in 1917.

Primates and Research Methods

What price did apes pay for this turn the scientific disciplines took, increasingly caught up in the quest for "objectivity" and "precision"? In the context of the time between the two world wars, followed by a period of struggle against fascism, then the Cold War, all possible experiments were justified; only the results mattered. Researchers were encouraged to do whatever was technically in their power to get the most profit from all possible resources. Primates constituted one of these resources. The orientation of scientific studies toward living things legitimated certain invasive procedures and a rigid respect for rules, sometimes transforming researchers into pitiless executioners. They were torn between two poles: on the one hand, the objective scientist free of all emotion; on the other, the sensitive human endowed with empathy. Some commentators, such as Donna Haraway, argue that these scientists' practices amounted to nothing more than sadism.[7] These experimental procedures locked animals and researchers into a framework where anything not open to measure and verification was ignored.

When tested in this manner, apes and monkeys were reduced to being mere *materials* of experimentation, or even martyrs to science. At the root of these studies was the boundary between humans and animals as well as the dichotomies of normal/pathological and scientist/amateur. Having gained scientific legitimacy, primatological studies were used to justify certain societal choices. The experiments on the attachment between mother and child done by Harry Harlow from 1957 to 1963 are a famous example. By separating young rhesus macaques from their mothers and exacerbating the experience of social isolation, scientists turned the animals into asocial and depressed beings, and by extension unintentionally stigmatized women who worked. Affirming that they had "made nature speak," scientists produced unimpeachable social norms by means of experimental procedures whereby apes were turned into guinea pigs.

The Baboons of Solly Zuckerman

Born in South Africa, Baron Solly Zuckerman, a zoologist, also supported strict experimental methods. In 1928, he studied a group of hamadryas baboons at the London Zoo. Zuckerman wanted to define what constituted a primate society, and his broader objective was to determine what rules would guarantee the reliability of observations. In this regard, he thought that only work done on animals in captivity was sufficiently rigorous. The hamadryas baboons showed themselves to be excellent examples of the "struggle for survival" proposed by Darwin: competition, aggression, and deadly attacks marked the daily life of the group. Based on these studies, which would, surprisingly, stand for all primate societies, Zuckerman declared that communities of primates were characterized by a strong hierarchy and male dominance.

For Zuckerman, this social organization was driven by various physiological mechanisms tied to reproduction. Zuckerman believed his study was perfectly scientific and rigorous, although he never took into account the fact that the population density of his study group was very high and its sex ratio completely unbalanced. His conclusions, which fixed the identity of baboons and the nature of the relationship between the sexes among primates, long dominated the discipline of primatology, confirming that the species constituted a good model to support the idea of an origin tied to the savanna. His work would be disproven by a number of later studies.

Zuckerman's later experiments are somewhat less well known. These studies give an idea of what scientists at the time were capable of and the multiple forms of complicity between science, industrial culture, and the military-industrial complex. Zuckerman is but one example of a researcher in a lab coat engaged in these practices. In the 1940s he tested the effects of bomb blasts on primates as well as their reaction when they were hit by different projectiles. In another study primates were struck on the head with a hammer in order to cause fractures so that the consequences of such impacts on soft tissue and bone could be examined. Since these studies were to be applicable to soldiers, it was important to conduct the experiments on animals similar to humans. The anatomical and primatological expertise of Zuckerman was thus brought into the military realm.

Rhesus macaques had already paid a heavy price in these kinds of studies. Starting in 1874, they were subjected to similar types of experiments. They were given, for example, repeated light blows on the head, death occurring without any apparent lesions. Later, their internal injuries were examined. For decades hundreds of individuals would endure the same torture, with variations consisting of the instruments used, species of primate, or targeted areas. The procedures were standardized, the phenomena measured, the range and scale perfected. Happily, not all researchers engaged in animal studies participated in these extreme versions of laboratory practices. However, a number of important figures in the scientific world worked on programs of applied psychology for the military-industrial complex. In the United States, Watson, Thorndike, Bingham, and Yerkes were, for example, employed by the Scott Company, which specialized in recruitment tests and troop motivation.

Primates Used in Medical Research

Since the beginning of the twentieth century, the number of experiments conducted on primates for biomedical and neurophysiological research has continually grown. Researchers have transformed primates into legitimate substitutes for humans without really questioning the pertinence of this model. The only similarities taken into account are anatomical and physiological. Other similarities, for example, the fact that animals suffer (not only physically but also psychologically), feel emotions, and remember traumatic experiences, are entirely ignored. The Cartesian concept of "animal-machine" hovers over all of this research as a necessary model, particularly in the field of experimental physiology, which, for example, supported the practice of vivisection. Known for his work on immunization and infectious diseases, Élie (Ilya Ilyich) Metchnikoff, a pioneer of modern Soviet experimental primatology, was among the first to use primates for biomedical research, which until then had employed dogs, guinea pigs, and rabbits. Starting in 1903, the naturalized French zoologist and bacteriologist developed a cure for syphilis by conducting experiments on primates.[8] Along with the German immunologist Paul Ehrlich, he received the Nobel Prize in Medicine in 1908. He used between fifty and sixty chimpanzees in his studies, thus paving the way for biomedical research using large numbers of apes.

A French "Monkey College" in Africa

In 1888 Louis Pasteur established his first institute in Paris. Several months later, he hired Metchnikoff, whom he named vice director in 1904. This promotion greatly influenced the research of the institute, especially in its orientation toward experiments using primates. Three years later, the first overseas Institut Pasteur was established in Saigon (today Ho Chi Minh City). Several other institutes would open their doors in Indochina as well as in Africa. In 1924, Pasteur set his sights on French Guinea and its chimpanzees: a primate colony was founded at Kindia by the bacteriologist Albert Calmette. It was a place that represented par excellence what Haraway calls "a fundamental part of the apparatus of colonial medicine."[9]

Built at the heart of a large tract of land at Kindia, Pastoria station housed eighty-nine chimpanzees and 106 monkeys in 1924. Over three years, approximately seven hundred chimpanzees were provided by African hunters. Nearly half of them died during the first two weeks after their arrival at Pastoria, even before they could be used in place or sent to Paris. To these deaths must be added the primates lost during their transfer from Africa to Europe and those who perished during the period of adaptation that followed. The institute was therefore equipped with a crematorium.

To the European and American public at large, however, it was presented as a "model village for primates," a "Monkey College," an "ape training school" where Westerners tried to "civilize" these uneducated primates (in much the same way they did indigenous populations). In this space, a colony within a colony, the primates were raised by African nurses while the whites "did science." However, the primates were not kept in a "nursery village" where life was good. Their enclosures resembled prisons more than houses designed especially for them. The courses said to "humanize" them existed only on paper: their only job was to become experimental material for biomedical research and studies in microbiology. The Bacillus Calmette–Guérin vaccine (BCG) for tuberculosis, for example, was tested at Kindia.

VIPs at Kindia

Eager to obtain funding, the directors of the institute at Kindia increased invitations to foreign researchers in the hope of establishing collaborations with Russia or the United States. But the station suffered from a lack of financing. The installations and the chimpanzees were neglected. Sent by the Soviet government and the Academy of Sciences, Ilya Ivanovich Ivanov was the first foreign researcher to go to Kindia. Calmette was greatly interested in artificial insemination, and Ivanov was the world expert. Calmette wanted to increase the number of chimpanzees available for research. There was no reproduction center at Kindia, and the number of primates in "reserve" was diminishing. Ivanov wanted to attempt experiments of hybridization between chimpanzees and humans, and inseminated three chimpanzees in 1927. Fortunately, however, he obtained no positive results.

Among the other visitors, an important researcher in the world of primatology arrived from the United States in 1929. Having established the

Yale Laboratories of Primate Biology, the American psychobiologist Robert M. Yerkes wanted to familiarize himself with the functioning of the station. His student Henry Wieghorst Nissen would remain in the field from February to March 1930, studying chimpanzee habits and assessing the feasibility of field research of chimpanzees. Nissen returned in June to the United States with sixteen chimpanzees destined to complete the quotas for the laboratories founded by Yerkes.

One of the essential missions of the institute at Kindia was to furnish the Institut Pasteur in Paris with primates. The other was medical experimentation. Apart from these experiments, the research of Albert Calmette, director of Pastoria, was limited to some measurements and a brief description of external physical characteristics, in the tradition of natural history practices.

Guillaume, Meyerson, and Nicole

Nevertheless, Calmette was eager to support different studies of primates, and at his request Paul Guillaume and Ignace Meyerson would attempt to explore the psychology of primates beginning in 1927. Using the particular approach of comparative Gestalt psychology, Guillaume and Meyerson conducted research on some thirty monkeys and apes at the Institut Pasteur in Paris as well as the Ménagerie at the Jardin des Plantes, which housed chimpanzees, orangutans, a gorilla, and gibbons. They tried to reproduce Köhler's experiments in the Canary Islands from 1912 to 1920, notably with a female chimpanzee named Nicole, whose intelligence, perseverance, and curiosity the two psychologists lauded. They also showed the importance of differences between individuals among primates. The orientation of their questions also allowed them to assess the limitations of their subjects and describe their shortcomings in comparison to humans, the "measure of all things."

The Institut Pasteur was not the only organization to set up laboratories in Africa. Located near Stanleyville (today Kisangani in the Democratic Republic of Congo), the Princess Astrid Institute was created in 1956 and officially opened in 1957. Biomedical experiments and the production of vaccines in vivo, principally the oral vaccine for paralytic poliomyelitis (or polio), were conducted by Hilary Koprowski in competition with Albert B. Sabin, who experimented on macaques. It was a new generation of laboratory, built on a large scale. It was equipped with what was called a "chimpanzerie" in the forest at the Lindi camp, located at the mouth of the river of the same name. Pygmy hunters brought their catch, killing members of a troop in order to capture a mother and her infant.

Between four hundred and six hundred chimpanzees would die in the laboratory at Stanleyville from its opening to its closure during Congolese independence in 1960. Within the framework of this research, chimpanzees were virtually denied the status of living beings. They were nothing more than organs and organisms for experimentation. The lives, habits, and social organization of the chimpanzees mattered not at all, as studying these would be of no utility. On the contrary: seeing apes as individuals would be detrimental to the research, which was entirely dedicated to scientific results, short-term profit, and productivity.

Endangered Species and Biomedical Research

In the United States, the first laboratory to use primates was started by Yerkes at Yale University, but the research conducted there did not focus solely on medical experiments. It was above all a laboratory of psychobiology. Beginning in the 1930s, psychologists and neurologists promoted the use of primates in scientific research. The need for primates grew. In 1935, the biologist Gertrude van Wagenen (one of the few women in the field of primate studies before the 1950s) founded a colony of rhesus macaques for use in biomedical research at the Yale School of Medicine. This species would quickly become one of the most often used in laboratories: they were easy to get from India and they were accustomed to people. Their immune systems and menstrual cycles were similar to those of humans.

John Farquhar Fulton created the first laboratory for neurophysiological research using primates as models at the same institution. In 1937 he delivered an important presentation ("The Chimpanzee in Experimental Medicine") that emphasized the advantages of research conducted on chimpanzees. Fulton was also one of the few researchers to conduct experiments on gorillas. Among the great apes, chimpanzees were the ones who would pay the most for medical research. Gorillas, being more rare, were too costly. In the 1940s, studies using chimpanzees increased even more.

At the time, polio was a devastating disease. Research into polio was conducted in a number of laboratories, including the Yale Laboratories of Primate Biology. In 1950, Hilary Koprowski developed a prototype for a vaccine. Two years later, Jonas Salk discovered a vaccine that could be injected, thanks to experiments using macaques. It was one of the greatest successes of modern medicine. In 1957, the Princess Astrid Institute began to produce anti-polio vaccines on a large scale for Koprowski, using chimpanzees. One year later, Albert Sabin, in competition with him, took control of the market with his oral polio vaccine of the same type. Polio was eradicated in the United States in 1994, and eliminated in many other countries during the 2000s.

The medical progress obtained by using great apes in the 1950s galvanized this type of biomedical research. Laboratories using primates grew, and the U.S. Congress began to finance these centers in 1960. The Animal Welfare Act was promulgated by the National Institutes of Health (NIH) in 1966. It would be amended seven times. Numerous violations of the law were nonetheless recorded. These laboratories increased in the 1970s thanks to NIH-funded research. In 1973, the Endangered Species Act was passed. It forbade the importation of chimpanzees captured in the wild. Laboratories took the necessary measures to continue their experiments: they developed breeding programs.

Monkey Business

The market for laboratory primates flourished: the subsidies went from less than $10 million in 1972 to more than $100 million in 1998. Chimpanzees were used as organ donors and crash test dummies as well as in studies of trauma, head crashes, windblast effects, and space program tests. (In experiments conducted by NASA in Project Whoosh, for example, chimpanzees were ejected from aircraft at supersonic speeds. None survived.)

They were used for biopsies, reproductive studies, drug testing, neurological research, injections into the brain, social isolation studies, and partial lobotomies. They were subjected to various kinds radiation to test their effects. They were inoculated with all possible infectious agents: HIV, polio, hepatitis, malaria, and so on. In 1974 Frederick Wiseman made a film about his experiences at the Yerkes National Primate Research Center. This film, *Primate,* vividly shows, without any commentary, the treatment inflicted upon the animals there.

During the 1980s, chimpanzees were considered excellent models for experimentation on treatments for AIDS until it was discovered that they were healthy virus carriers. Nevertheless, researchers continued to test the toxicity of new medicines on hundreds of chimpanzees, looking for cures for hepatitis B and C. In 1990, eleven North American laboratories were still using approximately 1,500 chimpanzees. At the time, 650 individuals were at the Coulston Foundation in New Mexico. The Laboratory for

The chimpanzee David in a cage at the Coulston Foundation (Alamagordo, New Mexico). Thanks to the extraordinary commitment of Carole Noon and her team, the Coulston chimpanzees are now housed in the Florida sanctuary Save the Chimps.
(Photograph courtesy of Save the Chimps)

Experimental Medicine and Surgery in Primates in northern New York (LEMSIP, New York University, 1965–97) had 300 chimpanzees and 300 other primates of different species. The funds allocated by NIH began to diminish in 1998. Four years later, the Federal Sanctuary System allowed chimpanzees held captive in different laboratories to be placed in sanctuaries where they could benefit from more dignified conditions of existence.

In 2006, 62,315 apes and monkeys (including approximately 1,500 chimpanzees) were still being used for medical and behavioral experiments in the United States. In 2010, the European Community banned all medical research using great apes, although 10,000 monkeys (out of 100,000 tested each year worldwide) are still used for medical research in Europe. The same year, the NIH asked a commission of experts at the Institute of Medicine for a report on the necessity of pursuing such research. These experts estimated that the majority of the studies were no longer necessary. The Chimpanzee Health Improvement, Maintenance, and Protection Act (CHIMP) was passed in 2013 to improve the daily life of primates. The six last American biomedical laboratories, nevertheless, had around 700 chimpanzees in 2015 when the U.S. Fish and Wildlife Service extended to captive chimpanzees the law protecting endangered species (Endangered Species Act), which until then had applied only to chimpanzees living in the wild.[10] Taking effect in September 2015, this law, based on New Zealand's Animal Welfare Act of 1999, constitutes a major advance for the well-being of great apes in the United States. Crowning the efforts of a number of associations and famous primatologists (including the indefatigable Jane Goodall), this law requires permits for keeping apes, drastically limits their use in scientific research, and protects them from any harmful treatment.[11]

In November 2015, the NIH stopped financing invasive research conducted on chimpanzees. The fifty chimpanzees kept for future research were placed in retirement. In April of the same year, the Nonhuman Rights Project filed a writ of habeas corpus (which examines the legality of a detention based on the right of personal self-determination) in the name of Hercules and Leo, the only chimpanzees still kept at the State University of New York at Stony Brook. For several hours, the judge, Barbara Jaffe, granted them habeas corpus, giving them a fundamental right heretofore reserved strictly for humans. She revoked this right several hours later, however, due to the far-reaching consequences this decision would have entailed. Many months later, judges denied the appeal of the animal rights associations. Hercules and Leo returned to the research center that had sent them to Stony Brook, the New Iberia Research Center in Louisiana. However, because of new laws, they would no longer be used for research. Since then, the only authorized studies of great apes are those tied to the survival of the species.

The Mental Life of Primates

In parallel with these biomedical experiments and behaviorist studies, there were fortunately other types of research taking place involving primate behavior and intelligence, two hot topics at the time. In 1916, in *The Mental Life of Monkeys and Apes,* Yerkes surveyed the existing studies of the "mental life" of primates. He cited the research of several experimental psychologists who, contrary to the behaviorists, accorded primates thought, including Möbius (1867), Hobhouse (1901), Sokolowski (1908), Haggerty (1909), Köhler (1917), and Pfungst (1912). He also evoked the work of three researchers who observed what he called "trained apes," those used to entertain the public: Hirschaff and his notes on the chimpanzee Consul (1905), Witmer on the chimpanzee Peter (1909), and Shepherd on two chimpanzees, Peter and Consul (1915; undoubtedly these were the same two chimpanzees). He ended by looking at studies more in the naturalist tradition: Cuvier (1810), Darwin (1859, 1873), Wallace (1869), Brehm (1873, 1888), Romanes (1900), Morgan (1906), and Holmes (1911).

Lightner Witmer did his research outside of laboratories. In 1909 he observed a chimpanzee, Peter, who performed in a music hall in Boston. Also described by the director of the Bronx Zoo, W. T. Hornaday, the ape skated, rode a bicycle, and smoked cigarettes. Witmer further affirmed that Peter perfectly understood what he was doing onstage, and that he liked doing it. The chimpanzee was also tested in a laboratory in the psychological clinic at the University of Pennsylvania. His performances were impressive: he understood what was being said to him, spoke the word *mama,* solved various problems, and imitated behaviors. In 1909 Witmer, a true visionary, predicted that in the future very young chimpanzees would be placed in the same conditions as young humans in order to be tested. As Dewsbury rightly points out, he thus anticipated the projects of home-rearing apes.[12]

Witmer also helped William Furness, who attempted to teach several words to an orangutan brought from Borneo in 1909. The ape was able to pronounce the word *cup* to ask for water. The orangutan was joined by a chimpanzee, acquired by Witmer, in 1911. During these experiments, the orangutan showed itself to be more patient, and thus easier to teach, than the chimpanzee. At the same time, George Gladden, assisting Hornaday at the New York Zoological Park, taught Suzie, a three-year-old chimpanzee brought back from French Congo (Congo-Brazzaville) by Garner, to respond adequately to forty commands. William Thomas Shepherd published an article on the mental processes of rhesus monkeys in 1910, then one on the intelligence of two trained chimpanzees in 1915, followed by observations of chimpanzees and orangutans at the National Zoo in Washington in 1923.

In Europe, Alexander Sokolowsky conducted research at the Hamburg-Stellingen Zoo, becoming the zoo's advocate in the scientific world. He published a number of articles between 1908 and 1923 on chimpanzees, gorillas, and orangutans. His work was criticized by Oskar Pfungst, one of the most famous European specialists of comparative psychology at the time and author of articles and books about the great apes at the Berlin Zoo (1912). Between 1911 and 1916 the Dutch naturalist Anton Portielje studied primates at the Amsterdam Zoo. He observed a female chimpanzee, Mafuca, who seemed quite different from the others. Portielje suspected that she might belong to a new species. Later, the chimpanzee was proven to be a bonobo: this was confirmed in 1963 after its remains had been examined by Harold Coolidge as well as by Peter Van Bree, who declared that it was undoubtedly the first bonobo to be kept at a zoo.

Psychologists and Apes

At the beginning of the twentieth century, research into primate behavior was limited to the developing fields of experimental biology and comparative psychology. Biomedical studies, however, still far outnumbered those of psychology. The first systematic long-term research of chimpanzee behavior and intelligence was undertaken by the German psychologist Wolfgang Köhler from December 1913 to May 1920 at a station created for this purpose on the island of Tenerife (the largest of the Canary Islands).

Five years later, Robert Yerkes founded the first American laboratory using primates at Yale University. Both of them subscribed to an experimental perspective as well as an evolutionist and continuist paradigm. Fascinated by great apes, the two researchers were interested in the species closest to humans as keys for understanding human behavior and mental processes. They both believed that the principles of "stimulus-response" and learning through trial and error proposed by the behaviorists were too simplistic and could not explain certain behaviors they classified as "intelligent." For them, these behaviors were based on much more complex processes; they believed the primates displayed a genuine capacity for reasoning. Köhler spoke of *Einsicht,* or *insight;* Yerkes of *ideational behavior,* or *ideas.* Their work would have an immense influence on the world of psychology as well as the budding field of primatology. Primates would also play an extremely important role in the development of the new discipline of comparative psychology.

The Chimpanzees of the Canary Islands

The first scientific station for primates in the world, the "Anthropoid Station for psychological and physiological research in chimpanzees and other apes," was announced in an article by Max Rothmann, a neurophysiologist at the University of Berlin, in 1912.[13] The Canary Islands were chosen for their climate and their relative proximity to Cameroon, at the time a German colony, where two subspecies of chimpanzees and one species of gorilla were found. The philosopher and psychologist Eugen Teuber was sent by the Royal Prussian Academy of Sciences in Berlin to build a research station for behavioral studies of great apes, the Prussian Anthropoid Research Center.

One adult female and six subadult (between four and six years old) chimpanzees were brought to the station: Sultan, Grande, Rana, Chica, Consul, Tercera, and Tschego.[14] Teuber made some observations but mainly confined himself to "training" the chimpanzees. Appointed by Carl Stumpf, his thesis director, Wolfgang Köhler arrived at the end of December in 1913 with his wife, Thekla Achenbach, and their two children, Claus (born in 1912) and Marianne (1913). Two other sons, Peter and Martin, were born at the station in 1915 and 1918. Although he began his work with little experience in animal psychology (apart from a course in psychology at Frankfurt), Köhler would direct the primate station and conduct a number of experiments, among them the most famous in the history of primatology, from 1914 to 1920.

Wolfgang Köhler, Sultan, and Grande

While at the time the dominant current of research into animal psychology was marked by the work of Thorndike, Köhler followed in the path of Hobhouse and drew inspiration from his book *Mind in Evolution* (1901). He believed Hobhouse's approach was very important, particularly because it accorded the capacity of "insight" to animals.[15] Both philosopher and psychologist, Köhler was influenced by Gestalt theory, which he helped to develop.[16] At the end of December 1913, he began his studies with the famous "Fruit Basket Experiment." The chimpanzee Sultan was presented with a basket of fruit suspended high up from a rope attached to a tree. After a moment of excitement, the chimpanzee seemed to pause, then in a flash understood the problem. Without proceeding through a phase of trial and error, he found a solution: he pulled on the rope to shake the basket holding the fruit, causing the delicacies to fall to the ground. Köhler had demonstrated that Sultan was able to determine a solution in an instance of "insight" after having mentally understood the different elements of the problem and establishing the relationship between these elements. This process was directly opposed to the trial-and-error model of learning, mechanical and random, proposed by Thorndike.

Köhler then conducted another experiment, this time with the female chimpanzee named Grande, who had to find a way to get to a banana placed far beyond her reach. Several boxes were dispersed on the ground. Grande stopped and concentrated, as if she were taking stock of the situation. All of a sudden, she seemed to understand what to do and began to place the boxes one atop the other. Perched upon this "tower," she got the fruit. She found the solution without any preliminary attempts, following a sequence of "pause-deliberation-solution." Thorndike's theories could not explain this type of behavior. Moreover, Köhler thought that Grande's action had all the more merit because, according to him, "wild" chimpanzees had no sense of static equilibrium. Determined to get the banana, Grande solved the problem through touch: a slight oscillation felt by the foot or hand led her to shift the boxes. She corrected the tower she had made using her touch and bodily sensations, as well as neuromuscular mechanisms, because she could not visually detect the imbalance. She showed herself to be skilled, careful, and persevering. Untiring, constantly verifying and readjusting her construction, she finished by climbing the boxes and attaining her goal.

Leaving aside for a moment his position as objective experimenter, Köhler greatly respected and admired Grande's tenacity. Chica, Rana, and Sultan were also able to attain the desired object. The others never managed. Köhler showed an individualization of behaviors and abilities: each individual found her or his own way of solving the problem. He also noted the use of sticks or pieces of straw covered with saliva to fish for ants ("like fishermen," he said) forty years before Jane Goodall, as well as the phenomenon of reconciliation among chimpanzees, later described by Frans de Waal.

Two Orangutans at the Prussian Anthropoid Research Center

Initially, Rothmann, the founder of the research station, had as his objective the comparative study of different species of great apes. During the First World War, two orangutans from Sumatra were sent to the Canaries. They arrived in June 1916. Köhler had never before seen any representatives of this species. He found the Asian apes very different from chimpanzees. He described the interactions between the two young primates, with the male (Filipe) being somewhat mistreated by the female (Catalina). He observed, however, a behavior of sharing food: Catalina offered an orange to Filipe. He also noted that the two young apes covered their heads with long leaves, like hats. Unfortunately, Filipe died of an intestinal illness shortly after his arrival.

Following some general observations (physical aspect, character, reactions, vocalizations, locomotion, and so on), Köhler studied Catalina's various behaviors. She was able to throw stones, carefully sighting her target, and in the same spontaneous fashion used sticks as weapons. Her enclosure lacking any structures for play, Catalina set about destroying it, with patience and tenacity, in a planned, nonrandom manner, using a stone as a hammer. Perfecting her demolition techniques, she often escaped from her enclosure. Köhler discerned a degree of insight in some of Catalina's actions. Using the researcher as a ladder, Catalina got up on Köhler's shoulders and directed him toward a piece of fruit that was out of her reach.

During another experiment, Catalina instantly understood the use she could make of a stick left at her disposition and got some fruit beyond her reach: as Köhler said, "The stick did become a tool."[17] She also sometimes made tools that she needed. She found some solutions on her own more quickly than the chimpanzees, although these were generally better, with Sultan remaining the best in Köhler's opinion. Catalina never piled up boxes, as did some of the chimpanzees. The psychologists thought that the morphological characteristics of orangutans, which are more arboreal than the chimpanzees, could explain the lesser performance. Yerkes, however, obtained similar results for the two species, though he was said to apply the same methodology as Köhler.

The Mentality of Apes

Köhler was particularly active during the six years he spent at Tenerife. An exceptional observer and open to a complexity ignored by the behaviorists, he took into account all the details and the subtle ways the chimpanzees perceived and managed their physical and social environment (somewhat like an ethologist *avant la lettre*). The author of numerous articles and books, he was the first to publish careful and detailed studies of the ability of chimpanzees to manipulate instruments and use tools by calling on their creativity and capacity for reasoning.[18] These works would galvanize research in cognitive psychology. However, Köhler was for a while slowed down in his work because of an accusation of espionage. According to the psychologist and writer Ronald Ley, the researcher may have been a communication agent for the Germans during his stay in the Canaries.[19] Ley does not produce any direct evidence but relies on numerous documents and testimonies. Köhler was asked to leave the station in 1918.

The decision to officially close the station came in July of the same year. The six chimpanzees living at the station (the chimpanzee Consul and the orangutan Catalina had died) were sent to Germany, arriving at the Berlin Zoo in mid-October 1920. The first director of the station, Eugen Teuber, often went to visit them, accompanied by his son Lukas. They were welcomed with calls of greeting by the chimpanzees. Even after more than seven years, the apes undoubtedly recognized a familiar face. Unfortunately, the chimpanzees did not long survive the German winter. The climate of the Canary Islands, the care taken by their keeper, Manuel González y Garcia, and the close ties they established with their human companions had apparently all contributed to their maintaining good health during their years on Tenerife.

The accusation of espionage and later his refusal to submit to Nazi dictates during World War II may explain why Köhler did not play the same role as founder of European primatology as Yerkes did in the United States, although he was in a position to do so. Köhler's departure for the United States in 1935 to escape Nazism as well as his desire to expand his work on Gestalt theory added to these reasons. The last founding representative of the Gestalt school (with Max Wertheimer and Kurt Koffka), Wolfgang Köhler died in New Hampshire on June 11, 1967. Heavily influenced by Gestalt theory, Köhler questioned the limits of other schools of thought in the field of animal psychology: to him, chimpanzees and orangutans had personalities, they thought, and they reasoned. It was a revolutionary viewpoint. His works on primate cognition, now classics, constitute a key stage in the history of studies of animal psychology, opening up an extremely fruitful field of research. They inspired many generations of primatologists, including Jane Goodall. *The Mentality of Apes* remains an essential part of the canon.

Apes as Guinea Pigs

At the beginning of the twentieth century, primates were not believed to be capable of reason. Interested above all in questions of learning, the behaviorists applied a simple stimulus-reaction schema to all behaviors and did not distinguish primates from other species. They did not observe: they recorded. They were interested in the performance of their animals (and it made no difference to them if these were rats, pigeons, or primates), which allowed them to formulate theories, principles, and general laws. Experimental subjects under this regime since the end of the nineteenth century, as well as standardized research objects in biomedical laboratories up until the beginning of the twenty-first, primates were subjected to procedures belonging to industrial culture, and sometimes even to constraints tied to the military-industrial complex.

Within this framework, young primates were taken from their mothers and raised by technicians in a standardized fashion in order to reduce the variations between individuals and to create a series of uniform organisms, ready to be tested. These primates were submitted to tests centered around human preoccupations; researchers pursued questions without any interest in the subjects, which were left with neither initiative nor any degree of liberty. They could exhibit only what the structures of the experiment allowed them to show. During biomedical or neurophysiological studies, they were even tortured.[20] They suffered physically and psychologically amid a general indifference to their fate. The apparent solidity of this way of doing science, as well as its placement within a framework of methodological rigor, long gave it prestige and authority in the eyes of the general public; the legitimacy of this type of research was rarely called into question.

Ayumu touching the numerals 1 to 9 in ascending order. Primate Research Institute, Kyoto University, December 13, 2006.
(Photograph courtesy of Tetsuro Matsuzawa)

FOUR

Great Apes in the Eyes of Scientists

What Does It Mean to Be an Ape?

VARIOUS DISCIPLINES, EAGER TO ESTABLISH themselves in the domain of behavioral research, were interested in primates: animal psychology, comparative psychology, psychobiology, behaviorism, and ethology. Initially studied in zoos, primates were later kept at research centers associated with these disciplines, all of them desirous of distinguishing themselves from "primate folklore" and later from "pop primatology," which were long considered by the public as reliable sources of information about primates.[1]

European ethology (the study of animal behavior), which was beginning to define itself as a discipline, paid little attention to primates during the 1930s and 1940s. This began to change, however, with the convergence of ethology and comparative psychology during the 1950s, the increasing recognition of primatology as a discipline in its own right, the number of primates available in the West, the work of dedicated researchers, the multiplication of possible fields of investigation, and the accumulation of knowledge about primates. All of these factors allowed for the gradual emergence of surprising observations about primates: reasoning; mathematical logic; language aptitude; self-consciousness; a sense of justice and morality; the ability to attribute mental states, reasoning, or knowledge to others; and a capacity to manipulate and trick their partners.

Robert Yerkes, the Father of American Primatology

Robert M. Yerkes played a major role in the emergence of primatology as a discipline in the United States as well as in the construction of the solid foundations upon which specialists would come to rely. A pioneer in the field of comparative psychology and the study of animal behavior, he spared no effort to promote a specific branch among these disciplines: experimental primatology.[2] To that end, Yerkes established the first laboratory for experimentation using primates in the United States. He then created one of the first scientific stations for great ape reproduction in the world.

At a time when travel was still uncomfortable and risky, he went anywhere primates were found, whether in Africa, Europe, Russia, Cuba, or the United States. Scholars as yet possessed little knowledge of the great apes, there being very few of them in the West. Yerkes lost no opportunity to find out more. Although he was entirely committed to a scientific and experimental version of primatology, he collaborated with "amateurs" (he would visit primates in private homes or circuses) and increased research outside of laboratories, those places of experimental science par excellence.

Most notably, he made contact with the owner of the largest primate colony in the world at the time, Rosalía Abreu, maintaining a correspondence with her that began in 1915 and lasted more than fifteen years. Her knowledge of primates interested him: few were then capable of keeping great apes alive throughout their normal life span. Abreu shared her notes and experience with Yerkes and invited him to come to Cuba. She even proposed founding a research station in Havana, promising to provide him with both primates and financial support for research. World War I prevented him from taking up this invitation immediately. His visit to Cuba would not come until much later, in 1924.

Untiring Traveler, Passionate Psychologist, and Practical Scientist

Yerkes established the first laboratory of experimental psychology using animals, the Harvard Psychological Laboratory, in Cambridge, Massachusetts, where he met the scientist Ada Watterson. They married in 1905. He became interested in primates in 1910. During a sabbatical he took in 1915, Yerkes made an in-depth study of the behavior of a young orangutan named Julius (as well as a dozen monkeys) at Montecito, California, thanks to G. W. Hamilton. One of his former students, Hamilton had worked on the sexual behavior of various species of primates, including baboons. At the time, Yerkes had already written on the maternal instinct of monkeys. The following year, Yerkes would officially present his project for a laboratory specifically dedicated to primate research after having published a monograph in 1916, *The Mental Life of Monkeys and Apes.*

By officially launching the idea of a well-defined program of experimental, systematic, continuous, and long-term research on primates under controlled conditions, Yerkes set in motion the professionalization of primatology as well as its recognition as an autonomous and legitimate discipline

Dr. Robert M. Yerkes with the chimpanzee Panzee and the bonobo Chim. (Photograph © Yerkes National Primate Research Center)

in the United States.[3] He declared it to be "by far the most important task for our generation of biologists."[4] He did not, however, acquire his first two great apes until August 1923: a male, Chim, and a female, Panzee, both three or four years old. Using his own money, he paid $2,000. This rather high price can be explained by the rarity of Chim, in reality a bonobo. Although not aware of this, Yerkes found Chim so much different from Panzee that he believed the young ape might belong to a different species.

In his view, Chim was something of a "genius" among apes. For Yerkes, fascinated by the intelligence of his student, his thirst for knowledge, his ability to adapt to his urban environment, and his inclination to appropriate certain human behavior, Chim became the prodigal son, the student all teachers dream of. After having observed the two primates at his summer home in Franklin, New Hampshire (bought in 1911 and named Franklin Field Station), Yerkes returned to Washington, DC, in September. The two young apes were installed in the third floor of his house there, with interior and exterior cages. Panzee died a few months later of intestinal tuberculosis.

Chimpanzees as Servants of Science

In June 1924 Yerkes left to visit Rosalía Abreu in Cuba, taking Chim, his secretary, and several scientific assistants. He stayed for a number of weeks at Abreu's residence (she was sixty-two at the time) and collected all possible information about the colony. A pioneer in long-term observations of chimpanzees, orangutans, and other captive primates, she had acquired her first primate, a macaque, in the 1890s and her first chimpanzee in 1902. By the time of Yerkes's visit, the colony comprised fourteen chimpanzees, three orangutans, gibbons, and many species of monkeys, including macaques, baboons, and mandrills. According to Wynne, she never had any gorillas.[5] Chim, stricken by a "respiratory epidemic" raging in the colony, fell ill, and died a month later of tuberculosis.

The same year, Yerkes joined the newly created Institute of Psychology at the institution that would finally allow him to realize his projects, Yale University. He opened the first university laboratory using primates in the United States, the Primate Laboratory of Yale University, part of the Institute of Psychology, in 1924, and finally succeeded in making chimpanzees "servants of science."[6] At the time, great apes were not much used in experimental research: they were rare, extremely costly, and fragile. They were also dangerous and unpredictable, especially as they got older. At the beginning, the laboratory had only two male and two female chimpanzees (named "the experimental group" by Yerkes): Bill, Dwina, Pan, and Wendy. Yerkes and his student Harold Bingham observed them both in the laboratory and at Franklin.

In 1925 Yerkes published two books on primates: *Chimpanzee Intelligence and Its Vocal Expressions,* based on his observations of Chim and Panzee, with Blanche W. Learned, and *Almost Human,* based in large part on the work and experiences of Rosalía Abreu, Alyse Cunningham (who had kept gorillas at her home in London), and the Russian scientist Nadia Kohts, who had been the first to conduct a comparative study of the behavior of a chimpanzee and a human infant. However, Yerkes had not yet visited Kohts in Moscow; he based his writings on information sent by the scientist. In 1929 the Rockefeller Foundation granted $500,000 to the laboratory (at the beginning, Yerkes's laboratory at Harvard had functioned on $500 per year). The small colony began to expand. Nearly ninety chimpanzees would be studied at the laboratory during the period that Yerkes was director.

Julius, Panzee, Chim, Bill, Dwina, Pan, Wendy, and Congo

From 1926 to 1928, Yerkes conducted the first systematic scientific study of the psychology and cognition of a mountain gorilla, captured in Congo by Ben Burbridge, an explorer and hunter. For three successive winters he stayed six to eight weeks in Jacksonville, Florida (at the home of Burbridge's brother and sister-in-law) in order to observe the behavior of Miss Congo. He carefully described the young gorilla, affectionately calling her a "gorilla-child" and describing their relationship as "friendly." He also organized experiments to test her intelligence. This study would call into question the ferocity and bestiality heretofore attributed to gorillas, about which, in reality, practically nothing was known. The psychologist would continue his observations of Miss Congo even after her transfer to the property of the Ringlings (the Ca' d'Zan) at Sarasota, the winter headquarters of the circus. He published a memoir, *The Mind of a Gorilla,* about the gorilla in 1927 (parts 1 and 2) and 1928 (part 3).

Yerkes was at the time the only person in the world to have scientifically observed and done systematic studies of the four species of great apes called "superior," with the orangutan Julius, the chimpanzee Panzee, the bonobo Chim, and the gorilla Congo, as well as the four chimpanzees at his laboratory. He attributed to them intelligence, a capacity for reasoning, and the beginnings of conscience and morality. Yerkes was the greatest primate specialist in the world. In 1929 he traveled to Moscow to meet Nadia Kohts. He also went to Africa to visit the Institut Pastoria at Kindia. The same year, Yerkes and his wife published *The Great Apes: A Study of Anthropoid Life,* a foundational work of more than six hundred pages. The two authors summarized all that was then known about the great apes as well as the research already done by Yerkes.

In June 1930, a site for primate reproduction, the Anthropoid Experiment Station of Yale University, was opened in Clay County, near the village of Orange Park fifteen miles southwest of Jacksonville.[7] With this breeding station, Yerkes reached another of his goals: to furnish enough primates for different programs of experimental research. Bill, Dwina, Pan, and Wendy were transferred to Florida over the summer. They formed the first group of breeders. A further thirteen chimpanzees from the colony of Rosalía Abreu, who died in 1930, were added, as well as sixteen other chimpanzees (six destined for the laboratory in New Haven) from Pastoria, thanks to Henry Nissen. In 1935, the colony comprised some thirty great apes, well acclimated and in good health.

The same year, a big event occurred at Orange Park: for the first time, a baby chimpanzee was born at a university laboratory. The daughter of Pan and Dwina (who died shortly after the birth of a postpartum infection), Alpha was born September 11, 1930. At the time, only seven chimpanzees had been born in captivity, four of them at the colony of Rosalía Abreu. Alpha grew up under the eyes of researchers. No chimpanzee had ever been more closely studied scientifically. The colony continued to develop. There were thirty-six births between 1930 and 1940, including a set of twins. Yerkes wanted to round out this program, already comprising two laboratories with field research. He sent his best students into the field to conduct a number of feasibility studies: Harold C. Bingham (gorillas), Henry W. Nissen (chimpanzees), and Clarence R. Carpenter (howler monkeys). He believed that "natural" behaviors and social organization should be studied in the primates' natural environment. In 1943 he published the second of his emblematic works, *Chimpanzees: A Laboratory Colony.*

Apes for the Future of Humanity

At the time, societies that encouraged research in eugenics were widespread. Yerkes believed that genetic research and selective procedures could benefit humanity. At this time of a crisis of civilization, there arose the desire to create a new human, modeled on scientific and technical innovations. Called by Yerkes "this little caricature of man," apes became the ideal and docile model for his project of the perfection of humans: if apes could be improved, rising from instinct to consciousness, humans could certainly be made greater. The apes were forced to become the brilliant students of this new space that was the experimental laboratory.[8] Yerkes retired in 1944 after twenty years of research. Nonetheless, as professor emeritus, he continued to work with the same vigor on various publications until he became ill in 1954. He died in New Haven on February 3, 1956.

Nadia Kohts and Joni

Research on great apes was also taking place in Russia. While Wolfgang Köhler was the first to conduct systematic studies at a research station in the Canary Islands, it was a woman, the Russian biologist Nadezhda Nikolaevna Ladygina-Kohts, better known as Nadia Kohts, who established the second site for experimentation on primates at the Darwinian Museum's Department of Zoopsychology in Moscow. She was, along with Köhler and Yerkes, one of the preeminent figures in primate psychology

Nadia Kohts and the young chimpanzee Joni.
(Photograph courtesy of the State Darwin Museum, Moscow)

at the beginning of the twentieth century. A student at Moscow Advanced Courses for Women, the young scientist worked in the State Museum, founded in 1907 and directed from 1911 to 1964 by her future husband, Alexander Fiodorovich Kohts.

From 1913 to 1916, she raised a young chimpanzee, Joni. However, he was often isolated in his cage and did not share his daily life with Nadia. His environment was thus very impoverished in relation to that of a young primate raised by its mother. This made the time he spent with Kohts, the closest thing he had to a mother, even more intense. Kohts observed Joni for two and a half years. She carefully described his general characteristics, habits, games, capacity for imitation, cognitive abilities, use of tools, empathy, and emotions. She experimentally demonstrated that the chimpanzee could see in colors. Joni failed the mirror test, most likely because he was too young. Kohts also studied his ability to identify forms and volumes, evaluate size, and recognize objects (hidden in a sack) by touch. She noted that he was able to draw and tie knots.

Following Darwin in *The Expression of the Emotions in Man and Animals,* she documented the entire range of Joni's facial expressions with dozens of photographs. Having escaped the preconceptions of Western researchers through her relative isolation in Russia, the young biologist showed herself to be full of empathy and sensitivity toward the young chimpanzee.[9] Kohts was very inventive in the design of her experiments, and she spoke of Joni's thoughts, emotions, and aesthetic sensibility. Having established a very close (thought by some to be even a bit too close) relationship with Joni, Kohts was in a position to gain a unique understanding of the chimpanzee. She created a complete, extremely detailed profile of the primate, who unfortunately died in 1916 of pneumonia at the age of four and a half. At the time, little was known about the care of great apes in captivity, and Joni undoubtedly would have benefited from better conditions. Kohts also observed a number of macaques and studied their capacity for imitation.

Infant Chimpanzee and Human Child

Nadia Kohts received her degree in comparative psychology in 1917, the year in which she established the psychological laboratory at the museum. Eight years later, she compared the notes and photographs she had of Joni with the data she had collected on her own son, Roody, since his birth. Although the lives of the two infants were in no way comparable, it was nonetheless the first study of comparative psychology between humans and great apes. In fact, Joni and Roody "met" each other only once, when Nadia brought Roody to see the mounted specimen of Joni and allowed her son to sit astride the chimpanzee for a photograph.

Robert Yerkes visited the laboratory directed by Kohts in 1929; their correspondence would last more than twenty years. In 1930 Kohts

Roody, Nadia Kohts's son, sitting astride the mounted specimen of Joni. (Photograph courtesy of the State Darwin Museum, Moscow)

published *Infant Chimpanzee and Human Child.* Translations of her writings were soon made in French and German, although the first English translation of her major work, edited by Frans de Waal, was not published until 2002. She was often criticized for her closeness to Joni as well as for her research topics and methods. The defenders of a rigorous experimental psychology treated her as an "amateur" and believed that her reputation as the pioneer of comparative psychological studies between humans and apes undeserved. She was, however, recognized as such by several major primatologists, including Yerkes, de Waal, and Harry Harlow, a strict scientist whom she corresponded with throughout her life.

The Wild Dream of Human-Ape Hybridization

During the same period, primates were also involved in certain ideological and revolutionary debates. An expert in artificial insemination and reproductive physiology, Ilya Ivanovich Ivanov wanted, as early as 1910, to conduct experiments in hybridization between humans and great apes.[10] Serological research on the mixture of the blood of humans and great apes had shown how much closer primates were to humans than any other species. Ivanov believed that if their blood was so similar, hybridization should be possible.

In 1917 he obtained the support of the Bolsheviks, who saw in these experiments solid arguments to support the antireligious propaganda of the Communist Party and further its materialist vision of the world. Science and technology were thus seen as allies in the struggle against the power that the Orthodox Church held in Russia. Scientific knowledge was associated with the will to radically break with the former czarist empire. A hybridization between humans and apes would prove the proximity theorized by Darwin.

While European researchers could get apes from their colonies, their Russian counterparts found primates expensive and difficult to obtain. In 1926, Ivanov first went to the Institut Pasteur in Paris, whose director, Emile Roux, supported him and suggested he conduct his research at the station at Kindia.

Before leaving for French Guinea, Ivanov met Serge A. Voronoff, who was known for his antiaging techniques using grafts from the testicles or ovaries of primates. In June 1920, the French (of Russian origin) surgeon and endocrinologist grafted for the first time tissue taken from the scrotum of chimpanzees and baboons onto those of a man. He had his moment of fame by promising greater life expectancy and renewed sexual vigor to his rich patients. Voronoff attempted to establish a farm to raise chimpanzees in Menton, France. His research was ridiculed, only to be later restored as pioneering work in antiaging techniques. Ivanov and Voronoff tried but failed to establish a project of fertilization of the female chimpanzee Nora with human sperm.

Artificial Insemination and Interspecies Hybridization

Ivanov arrived at the Institut Pasteur in Kindia in 1926 and the following year inseminated three females from the colony with human sperm. None of the trials were successful. The primates, he thought, were simply too young. During a second voyage to Africa, Ivanov continued his research using older chimpanzees at Camayenne (Conakry, Republic of Guinea) at the Colonial Botanical Gardens. He performed a number of inseminations, using sperm from local men, principally on the chimpanzees Babette and Syvette. Aided by his son, he conducted these trials at night, in secret, under difficult conditions. None of the females inseminated became pregnant.

Ivanov returned to Russia with two monkeys and thirteen chimpanzees captured in Guinea. He went to the first genuine primate breeding station in the world (under the authority of the Soviet Institute of Experimental Pathology and Therapy), established at Sukhum in Georgia, on the shores of the Black Sea, in 1927. He wanted this time to experiment in fertilizing a human female with the sperm of a great ape. In 1928 he approached Rosalía Abreu, hoping to get a male ape from her private colony in Cuba. Initially somewhat favorable to the project, Abreu eventually refused, fearing the reactions of others. Ivanov then envisaged using the sperm of the orangutan Tarzan, at the time twenty-six years old and the only available ape at Sukhum, but he died in June 1929, before the experiment had even begun. Ivanov had, however, found a number of women who wanted to try the experiment.[11] But Ivanov, who had fallen into disgrace under Stalin, was arrested and sent to Alma-Ata (today Almaty, in Kazakhstan). Freed after five years of detention, he died in 1932 the day before his departure for Moscow. The details of his research are buried in the Soviet archives.

Biomedical research, however, continued at the Sukhum Primate Station, which also had a nursery. Supplying primates to over fifty research institutions, the station expanded greatly after World War II: it had 773 primates in 1957 and 2,018 in 1966. Various researchers conducted their studies there, as did the Soviet atomic and space programs. Destroyed during the battles for independence in 1992, the station was moved to the

Sochi-Adler region (across the border in southern Russia) under the name of the Scientific Research Institute of Medical Primatology. There is, however, still a center at Sukhumi (as it is now known), the Abkhazian Primatological Research Center of Experimental Medicine.

Ivanov's project, falling within the realm of eugenics, was to enlist apes in the service of the state in order to create a new human. Yerkes supported a similar project around the same time in the United States, the goal of the two researchers being to select a series of traits conforming to an ideology then espoused in their countries.

Ivan Pavlov's Primate Research

The work of Ivan Petrovich Pavlov (Nobel Prize winner in Physiology or Medicine, 1904) on dogs and his studies of conditional reflexes, one of the major areas of experimental psychology at the time, is well known.[12] Less well known is his research on great apes.[13] However, these experiments had a major influence on the great debate between behaviorism and Gestalt theory.

Pavlov visited Yerkes's laboratory of experimental psychology in New Haven in 1929, during the Ninth International Congress of Psychology at Yale University. The two men remained in contact and exchanged publications. Pavlov sent Fursikov, Ivanov-Smolensky, and Rait-Kovaleva to the Leningrad (today Saint Petersburg) Zoo. They conducted research there on primates until 1926. In addition, a collaboration was established with the Sukhum Primate Station founded one year later. A former student of Pavlov and assistant to the director of the station, Leonid Voskresensky, sent Pavlov film and reports of his experiments on the use of tools among baboons, chimpanzees, and orangutans. Some of Pavlov's collaborators went there to test his theories on conditional reflexes with primates. Comparisons could then be established between the behaviors of dogs, monkeys, apes, and humans. Later, Pavlov sent another of his students, Petr K. Denisov, to visit Voronoff's laboratory in the south of France. These various experiences show how much Pavlov was already interested in primate studies.

At the end of August 1933, Denisov returned to the Koltushi Biological Station near Leningrad with two chimpanzees he had received as gifts. This station was conceived as a veritable "village of science" by Pavlov. Showing a great deal of curiosity about Rafael (six or seven years old) and Roza (about five years old), Pavlov began some preliminary observations and experiments, including crossing water using a pole, cooperative problem solving, pyramid building, and stacking boxes. In 1934, Lev Vygotsky published a major work in Soviet psychology, *Thought and Language,* in which he described at length the work of Köhler and Yerkes on great apes.[14] Pavlov, however, already knew of the work of the German psychologist. He had visited his laboratory at the Psychological Institute in Berlin at the end of the 1920s (probably 1927), shortly after Köhler's return from the Canary Islands. He also participated in the psychology congress held at Yale in 1929 in the same session as Köhler. He had read Köhler's books *The Mentality of Apes* and *Gestalt Psychology* by the time he was conducting experiments on Rafael and Roza, which were systematically filmed beginning in the spring of 1934.

Rafael and Roza, Pavlov's Chimpanzees

Working for the first time with primates, Pavlov repeated certain of Köhler's experiments with the chimpanzees Rafael and Roza. Assisted by Denisov, he questioned the holistic approach and notion of insight proposed by Köhler. Although Pavlov and Denisov obtained the same results as Köhler, they did not explain them in the same way. From their point of view, the chimpanzee arrived at a solution by the accumulation of perceptions, which led it to reinforce or inhibit certain behaviors. In the course of this process, the ape established associations and acquired progressively diverse conditioned reflexes.

For Pavlov, the moment of pausing and reconfiguration of the situation, the "Köhler moment," was in reality a sign of fatigue, accompanied by a lowering of the investigatory conditional reflexes that followed, logically, a period of excitation. The vision of a reward stimulated the brain of the chimpanzee that then, newly under a cortical excitation, set its efforts toward obtaining the food. Pavlov described the process as a chain of reflexes, linked to a schema of stimulus-reactions within a framework of trial-and-error processes, not as an entirety of sequences of significant events in relation with rich interactions between the organism and its environment. He categorically rejected Köhler's idea of problem solving through insight, and criticized his mentalist and subjectivist approach.

The final phase of experimentation consisted of combining different problems that the chimpanzees had to solve one after the other in order to arrive at a solution. Pavlov believed that although great apes attained very high cognitive abilities, they differed from humans, principally by reason of language, which allowed humans to achieve even more connections and associations. However, certain things specific to chimpanzees, for example, the fact that they have four "hands," allowed them to achieve complex associations. He further undertook the editing of two manuscripts to gather his reflections: "The Intellect of Anthropoid Apes" and "Psychology as a Science."

Pavlov's studies of chimpanzees were also, for him, the place of the confrontation between the neuropsychological tradition proper to *experimenters* (who isolated the animal and certain measurable elements of its behavior to the laboratory, cutting it off entirely from its world) and the European psychological tradition tainted by naturalism and tied to *observers* (more specifically Gestaltists) and between the paradigm of conditioned reflexes (close to associationism) and that of insight (tied to mentalism).

On February 21, 1936, Pavlov went to Koltushi, as he had every Friday since 1933 (he also spent summers there). He observed for one last time the two chimpanzees that were at the heart of a crucial debate between the two dominant currents of psychology at the time. He died a week later in Leningrad, on February 27; he who had worked so much with dogs ending his exceptional career with studies of great apes.

The Slow Emergence of a Discipline

Despite the accumulation of work done by a number of important scientists, the study of primates did not constitute an independent discipline.

Theodore Ruch first coined the term *primatology* in 1941, according to John Fulton in his preface to Ruch's *Bibliographia Primatologica: A Classified Bibliography of Primates Other Than Man.* Ruch gathered together most of the extant bibliographical references on anthropoid and nonanthropoid primates since Aristotle. In his title, he used the Linnaean denomination "primate," taking from it the name of a new discipline: primatology. The overwhelming majority of topics are studies of anatomy, osteology, embryology, physiology, or work on the functioning of different systems: reproductive, circulatory, nervous, sensory, digestive. At most, only 20 percent of the entries concern other topics: psychobiology, intelligence, memory, symbolic abilities, learning, imitation, sexual life, cooperation, learning through trial and error, and even art or mythology. Several references treat the question of the origins and possible convergences between humans, ancient hominids, and great apes as well as the place of humans among primates.

Leaving aside the fields of anatomy, physiology, medicine, and pathology, some researchers began scientifically to study the behaviors, mental states, and cognitive abilities of great apes. They did so carefully because the many resemblances between primates and humans posed a problem of principle. They risked placing in peril two of the pillars of applied experimental science on animals: the obligation to be objective and its corollary, the struggle against anthropomorphism. What is more, from the point of view of some biologists, the fact that "primatologists" focused on those species closest to humans showed that their interests were centered on humans and indicated an indifference to natural diversity. This placed primatology closer to the human sciences: primates constituted a means for thinking of humans, while biologists focused on animals.

Primatology and Ethology

While psychologists very quickly became interested in primates, this was not the case for the first European ethologists.[15] Coming from a naturalist tradition (notably represented by Réaumur, Buffon, Guer, Le Roy, Lamarck, Flourens, Georges and Frédéric Cuvier, Étienne and Isidore Geoffroy Saint-Hilaire, Darwin, Wallace, and Fabre), the ethologists engaged in rigorous scientific observations in a "modern" perspective of the study of animal behavior (then represented by Spalding, Whitman, Wheeler, Holmes, and Craig).[16] The first ethologists had as their objective the description of the specific behaviors of different species (either wild or domestic) in their environment. They focused mainly on the birds, fish, and insects native to their regions. They thought that ethology did not pertain to the study of primates or rats. In their minds, these species were closely linked with questions of learning, dear to comparative psychologists and behaviorists.

The pioneering ethologists were influenced by the work of Jakob Johann von Uexküll (Lorenz proclaimed himself to be his student), who gave animals the status of *subjects.* The German biologist and philosopher refused the reified version of animals proposed by behaviorists: he saw them as shaping their own world, using objects that made sense for them, in their own environment. In an attempt to understand the animals' point of view, he created the notion of *Umwelt* ("world" or "representative universe"). However, von Uexküll was more interested in ticks, moles, and jackdaws than in primates.

The Dutch biologist, anthropologist, and psychologist Frederik Jacobus Johannes Buytendijk helped to spread von Uexküll's Umwelt theory. Buytendijk subscribed to a physioethological and phenomenological approach to animal behavior, criticizing biologist reductionism and advocating the study of living beings in their natural surroundings. He also saw animals as subjects acting to operate a synthesis of the discrete events of the outer physical world, as reflected by their sensory-motor integration, thus distancing himself from the idea of a "chain of reflexes" supported, for example, by Pavlov.[17] The head of the school of existential-phenomenological psychology at Utrecht (the Netherlands), Buytendijk was interested, from the time of his thesis, in the formation of habits among animals, clearly showing the influence of Lorenz on his work. He also cited the work of Yerkes, Nissen, and Hayes. Buytendijk described the chimpanzees Gua and Viki as well as others raised in human families. He studied the facial mimicry of primates while avoiding any anthropomorphism and working in both a rigorous and original fashion. Noting the capacity of primates to adapt to the most diverse environments, Buytendijk stated that their behaviors were "close" to those of humans, and that relations between the two species could attain unequaled depths. He maintained, however, that the differences between humans and apes were fundamental. He argued that the appearance of humans was explained by the emergence of a new principle in the animal world tied to language, creative work, objective perception, and the liberty taken in regard to normative obligations.

A Synthesis of Ethology and Comparative Psychology

The Austrian Konrad Lorenz and Nikolaas Tinbergen from the Netherlands laid the foundations of ethology. Lorenz's mentors were Julian Sorell Huxley and Oskar August Heinroth; he considered the latter one of the founders of the comparative study of animal behavior as a branch of biology. The third founder of ethology, the Austrian zoologist Karl von Frisch, set forth the idea that bees were capable a form of communication (somewhat scandalous at the time), and published his work about their "dance" in 1927.[18]

Wanting to place ethology within the field of science, Lorenz gave it conceptual, methodological, and institutional bases during the 1930s. Between the two world wars, he collaborated with Tinbergen, who possessed solid laboratory experience. During the summers they worked together at the Lorenz family house at Altenberg, Germany, especially on the behavior of geese, which made Lorenz famous. They (along with Karl von Frisch, Heini Heidiger, and Irenäus Eibl-Eibesfeldt) founded cognitive ethology and placed it within a Darwinian perspective. Ethology and comparative psychology thus shared the field of studies of animal behavior. The European ethologists saw themselves in opposition to psychologists, specifically the American behaviorists (influenced by the work of Pavlov on classical conditioning) and their experimental designs. For Lorenz, behaviorists knew nothing of animals since they studied them in laboratories. The American researchers thought, on the other hand, that field research had no validity due to its lack of experimental controls. In 1938, however, Tinbergen visited several American laboratories, including both of Yerkes's, but later declared he felt ill at ease with Americans' methods and the goals they pursued.

After World War II, Tinbergen tried to reconstruct the discipline of ethology. The radical eugenicist positions Lorenz supported during the war had temporarily thrown a shadow over the discipline. In 1949, a large international conference, "Physiological Mechanisms in Animal Behavior," was organized at Cambridge. Here the fundamental concepts of ethology were redefined. During the 1950s, European ethology and American comparative psychology continued to clash. The two sides pitted instinct against learning; specific characteristics tied to each species (such as the intelligence of chimpanzees) against rules of learning generalizable to different species (learning through trial and error by rats, pigeons, dogs, and chimpanzees); and studies conducted in a natural environment against controlled experiments in the laboratory. In 1951, Tinbergen published an overview of the discipline, *The Study of Instinct.*

Ethology underwent a period of adaptation and change in the 1960s, a time of ascendancy for radical behaviorism and a glorious period for Skinner with his rats and mazes. Ethologists adopted new techniques for the collection and analysis of data, set forth different procedures, and examined the theoretical and conceptual bases of the discipline. In his article "On Aims and Methods of Ethology" (1963), Tinbergen posed four questions to be taken into account when studying animal behavior. Two of these concerned immediate causes (the "how") and the other two ultimate causes (the "why"). They would have a long-lasting influence on the discipline. In 1966 Robert Aubrey Hinde published *Animal Behaviour: A Synthesis of Ethology and Comparative Psychology.* With Gregory Bateson, he tried to minimize the differences between the two disciplines, stressing the complementarity of research done in a natural setting and experiments conducted in the laboratory, which allowed for the exploration of underlying mechanisms impossible to test in the field.

"Take That, Behaviorists!"

During the 1960s, ethology underwent an impressive development. Field primatologists were significantly influenced by ethological methods. They went into the field with notebook and pencil in hand, meticulously observing and describing primates, taking notes, and writing logbooks. One of Hinde's students at Cambridge, Jane Goodall, is an excellent example. Like Lorenz, she did not hesitate to habituate her study subjects to her presence, as well as to share parts of their lives with them, in order to approach and more closely observe them. Konrad Lorenz, Nikolaas Tinbergen, and Karl von Frisch received the Nobel Prize in Physiology or Medicine in 1973. None of them studied mammals (although Lorenz did have a capuchin monkey named Gloria).

Upon receiving the Nobel Prize, Lorenz later confessed to having thought: "Take that, behaviorists!"[19] Primarily zoologists, the objectivist ethologists had, however, gradually adopted the procedures of experimental science. During the 1980s their publications became more mathematical and quantifiable. Lorenz deplored this. In his eyes, these means were typically applied to inorganic matter. The ethograms (indexes of behavior) and sociograms (indexes of interactions between individuals), the various recordings and measurements (making use of increasingly sophisticated technologies), the field quadrants, the quantifications, tables, and statistics—these had taken the place of ad lib observations and notebooks. The discussions at the ends of articles became less and less detailed and more and more cautious: after all the effort spent collecting data, the scientists remained guarded about their interpretation. Primatological studies followed the same development.

A New Orientation of Primatological Practices

For a long time, primatology was not, properly speaking, a scientific discipline—that is to say, a discipline defined by its own questions, practices, objectives, methods, and instruments rather than only by its object of study. It was instead a field that brought together different disciplines around the same object: primates. However, the work of preeminent figures such as Yerkes and Köhler, the growing number of publications, and the emergence of areas suitable for the exclusive study of primates gradually changed this. In 1931 Harold J. Coolidge gave the first courses specifically on primatology (before it was known as such) at Harvard. Within museums, where the collection and description of specimens continued, biologists finished putting in order a still chaotic classification of primates.

Primatology gradually began to take on the traditional suite of things associated with proper scientific disciplines: scholarly societies, specialized journals and reading committees, conferences, academic titles, institutional support, and international exchanges. The first specialized international journal, *Primates,* was published in 1957. In the 1960s, Jane Goodall and Toshisada Nishida began long-term field research on great apes in Africa. Furthermore, experimental studies continued, opening up new areas of research. Seven centers of primatology were active in the United States. The International Primatological Society was founded in 1966, and the American Society of Primatologists in 1981. Their journals, the *International Journal of Primatology* and the *American Journal of Primatology,* were published for the first time in, respectively, 1979 and 1981.

When Genetics Is Brought into the Study of Primate Behavior

William Donald Hamilton, John Maynard Smith, and Robert Trivers integrated neo-Darwinist theories into the discipline of ethology. In response to the variability of behavior and the instability of social organization in certain societies (for example, that of primates), Hamilton proposed the idea that stability resided not in the social structures but in the genes. As the Swiss primatololgist Hans Kummer would later state, "the things individuals do to assure their descendance are nothing other than behaviors fashioned by their genes towards this end."[20] Selection favors genes that facilitate the most advantageous biological characteristics for survival.

Until this time, the concept of altruism constituted a flaw in evolutionist theories. How could this type of behavior (which in no way rewarded the altruistic individual) be explained in the Darwinian context of the "struggle for survival"? Hamilton expanded the question to the genes held in common by a related group. He developed the theory of global genetic success to justify altruistic behavior: aiding an individual who shares one's genes favors the diffusion and perpetuation of these common genes. Aligning certain behavior with reproductive success (or fitness), sociobiologists were in this way able to explain not only cooperation and altruism among the great apes but also cases of infanticide. With the goal of favoring their own descendents, male gorillas and chimpanzees relieved females of maternal obligations tied to the progeny of another male by killing their babies. They thus made the females sexually receptive more quickly.[21]

Popularizing sociobiology in a best seller, the American biologist and naturalist Edward Osborne Wilson wanted to extend sociobiological theories to all living things, including humans, which raised a great deal of debate.[22] Inspired by Lorenz, he placed animal behavior at the heart of his theories. While they reasonably covered the entire field of studies of animal behavior, ethology, behaviorism, and comparative psychology became more and more threatened by sociobiology. Specialists of the great apes turned toward a more experimental and quantifiable version of their discipline. Wilson predicted that ethology would be soon absorbed by behavioral ecology, while comparative psychology would be overtaken by integrative neurophysiology and sensory physiology. New questions took the place of the old: fitness, sex selection, evolutionary stable strategy, population genetics. The living sciences and behavioral studies changed their paradigm: it was now a question of taking them from the hands of comparative psychologists and giving them to the biologists. Within this framework apes were viewed from an accountant's perspective, as managers of time-energy budgets that optimalized a cost-benefit equilibrium. The individual was reduced to nothing but a vehicle, a temporary support, for his or her genetic heritage.

The Difficult Problem of Consciousness Among Apes

Other questions nevertheless continued to disturb the field of primatological studies. For a long time, research into thought, consciousness, and mind were excluded by a dominant current within comparative psychology, particularly due to the inaccessibility of mental processes and the fear of anthropomorphism. Post-Watsonian and post-Skinnerian behaviorism as well as experimental psychology took up this interdiction, taking into account only the observable aspects of behavior.

However, by the end of the 1950s certain intellectuals began what would later be called the "cognitive revolution" by Howard Gardner in *The Mind's New Science* (1985). Opposed to behaviorism and Skinner, Noam Chomsky affirmed that certain behaviors could provide data for internal mechanisms of the mind.[23] These mechanisms, the brain and its functioning, were slowly reinscribed as legitimate objects of study within psychology, biology, and ethology as part of a multidisciplinary approach. For Jerome Bruner, it was above all an attempt to put the process of "meaning-making," though here applied to animals, at the center of the psychological disciplines. This research focused on primates for thirty years. It would, moreover, open up the development of different fields of study, including cognitive science, artificial intelligence, computer science, and neuroscience.

In 1974, in a famous article about bats, the American philosopher Thomas Nagel pointed out the fact that even if the system of echolocation were objectively understood, one could never have an idea of the subjective experience of the animal: the qualitative dimension of this experience is in effect inaccessible to us. Nagel insisted on the limits of materialist studies: they could not account for the phenomenological aspect of consciousness.[24]

In response to the epistemological pessimism of Nagel, Donald Redfield Griffin said that we should not refuse to tackle a phenomenon simply because we had only a partial comprehension of it. Thus, two years before Premack and Woodruff's article, "Does the Chimpanzee Have a Theory of Mind?" (1978), Griffin established a program concerning this field of study, heretofore neglected, and officially established cognitive ethology, a synthesis between ethology and the cognitive sciences. In *The Question of Animal Awareness: Evolutionary Continuity of Mental Experience* (1976), he stated his desire to work in a scientific manner on the subjective experiences and thoughts of animals, on the "difficult problem of consciousness," seen as biological phenomena through an organ, the brain.[25] Believing that intra- and interspecies communications constituted a privileged mode of access to the mental life of animals, Griffin drew inspiration from the work of von Frisch on the communication of bees and Gardner's studies of the female chimpanzee Washoe.[26]

Cognitive Abilities Among the Great Apes

Griffin's work was criticized, particularly by William A. Mason (who himself worked on primates), for its various flaws: methodological errors, theoretic problems, confusion between mental representation and mental experience, and misperceptions between the capacity to anticipate future actions and conscious intention. It did, however, open up the field of studies to the psychic and cognitive lives of animals and proposed a different way to account for certain behaviors: "Cognitive ethology brings something very new, by according to each animal the possibility of being individually intelligent by best mobilizing the cognitive abilities of its species."[27] While disagreeing with certain of Griffin's propositions, Daniel Dennett, an American philosopher of science specializing in philosophy of the mind, offered a sort of philosophical legitimacy to these studies by proposing an empirical approach to consciousness that incorporates scientific data.[28] However, he believed that only humans are capable, because of their capacity for language, of "high-level" representations that characterize consciousness: the representation of representations or the representation of one's own consciousness. He excluded apes from this level of consciousness.

The first work done on the capacity of great apes to also have representations goes back to a "mentalist vision." It began with Walter S. Hunter in 1912.[29] Reprising these studies in the 1930s, and very interested in the work of Wolfgang Köhler, Otto Leif Tinklepaugh showed that chimpanzees were capable of imagining the space where they were held, as well as the significant elements it was composed of.[30] In 1932 Edward Chace Tolman put forth the ideas of a cognitive map and of the animal being an active subject in the research; these propositions were later taken up by the cognitive school, of which he was one of the precursors.[31] He also questioned the behaviorist conception of learning by association.

As did various other primate specialists, Emil Menzel worked both in situ and ex situ. He thought of himself, however, as a naturalist before an experimenter. After having done fieldwork on Japanese macaques in the 1960s, he studied the "strategy of the traveling salesman," or strategy of least distance, among great apes. He showed that they mentally possessed a spatial cartography that allowed them to organize their trips in search of food based on a model of Euclidean geometry. During the 1970s he notably tested the ability of chimpanzees to guide others to the precise location of a hidden object by pointing a finger from a distance.

Appearing at first glance simple, the transitional gesture of pointing is actually more complex. Revealing a process of joint attention, it is part of a threefold interaction (comprised of the one who points, the one who sees that gesture, and the object) and implies intersubjectivity, intentionality, communication with the goal of attaining an objective, and an invitation to cooperation. Emil Menzel's son, Charles R. Menzel, continued this type of study, particularly on spatial cognition, recall memory capabilities, and the prospection abilities of chimpanzees and bonobos. Further, some researchers show that primates are able to "communicate" about absent objects.

Chimpanzees Have a Mind of Their Own

In regard to questions of representation, another paradigm shift was initiated by the psychologists David Premack and Guy Woodruff, who formulated in 1978 a "theory of mind," proposed from studies of apes: they showed that chimpanzees possessed *certain* mental states and were capable of attributing them to others.[32] Chimpanzees can understand that others see (and thus take into account their spatial perspective), hear, and think certain things. They are able to imagine the emotions, perceptions, objectives, intentions, and knowledge of those around them, and respond with appropriate behaviors, or even by strategies, on this basis.

Many researchers, including Daniel Povinelli, refuted the possibility of a theory of mind among the primates. Revising his initial findings, Michael Tomasello affirmed the contrary and went even further: he thought that great apes were capable of relating the mental states they attributed to other individuals with the manner in which these individuals acted in a given situation. They were thus able to comprehend, in a subtle fashion, the psychic process linking intention, objectives, and actions in others, just as Cuvier had long ago noticed in Empress Joséphine's orangutan. In 1992, the primatologist Christophe Boesch assembled various data supporting a theory of mind among chimpanzees that he observed in the Taï forest.

Tomasello ended the controversy by producing sufficient data to affirm, against Povinelli, that the theory of mind did indeed apply to chimpanzees, and furthermore that it was not simply a superficial understanding of the behaviors of others. He admitted, however, that there was no proof of the ability of apes to understand that an individual had *false* beliefs about a given situation. It also seemed that chimpanzees were able to understand the *intentions* of others. That did not necessarily mean, however, that they could *share intentions* with others (the sharing of intentions perhaps constituting the keystone of human cultures). It appeared in any case that apes were able to apprehend much more complex mental states than other primates.

Researchers explored not only the ability of primates to imagine certain mental states of others, or *folk psychology* (the theory of mind and the guesser-and-knower paradigm of modern intersubjectivity), but also their ability to apprehend the physical world, or *folk physics* (the theory of causality), and to understand other species, or *folk biology*. Even though apes were tested in anthropocentric experimental designs far removed from their world and cut off from all social ties, various primatologists (such as Povinelli, Tomasello, Boesch, Call, and Hare) eventually showed how apes are able to understand their physical and social world and demonstrated their ability to develop varied strategies—to manipulate, lie, cheat, scheme, or voluntarily trick others with tactical intention—which constitutes a certain evolutionary advantage for these eminently social species.

However, some criticism emerged concerning the ensemble of studies of cognitive ethology because of their focus on primates, particularly chimpanzees. In 1982 Benjamin Beck denounced what he called chimpocentrism.[33] Taking the example of the behavior of some gulls of breaking open shells by dropping them from a height onto hard surfaces, he explained that concentration, determination, space-time representation, mental cartography, selectivity, learning by trial and error, and elaboration of strategies were not the sole domain of tool-using primates, nor even of mammals.

Complex cognitive abilities existed in other species that scientists did not pay enough attention to, essentially for reasons of anthropocentrism (great apes being the species closest to humans). According to Beck, the field of study of cognitive evolution therefore needed to be enlarged to take into account the diversity of intelligences.

Mirrors and Self-Consciousness

Inspired by Darwin's observations at the London Zoo, Gordon Gallup tested the ability of apes to recognize themselves in mirrors.[34] This kind of study would enjoy a certain popularity, particularly because of its supposed ties to the questions of self-consciousness and the aptitude to have a theory of mind. When a mirror was placed in its enclosure, the chimpanzee tested by Gallup first used it to look at parts of his body he could not examine otherwise (for example, the inside of his mouth). After he became used to the mirror over several days, the ape was anesthetized and a spot of color was painted somewhere, such as above his eye or on his ear. When he woke up and looked at himself in the mirror, he touched the spot with his fingers and tried to rub it off, which showed that he understood it was indeed his own body that appeared in the mirror, not that of another ape.

Experiments proved that particular animals were able to recognize themselves in mirrors: chimpanzees, bonobos, gorillas, orangutans, dolphins, orcas, magpies, pigs, and Asian elephants all showed this capability. However, could it be as easily affirmed that the fact of identifying one's reflection was proof of the existence of self-consciousness? One of the best French specialists in primate cognition, Jacques Vauclair, believed that there was confusion between *representation* of self and *consciousness* of self (self-awareness). All that could be said, he added, was that Gallup's chimpanzee was able to picture its own body.

However, it would seem that certain anecdotes reported about the "talking" apes, if true, would argue for their having a form of self-awareness.[35] For example, the female chimpanzee Viki, raised in a human family, classed her own photograph as belonging among snapshots of humans rather than photographs of animals. According to Francine Patterson, Koko, a female gorilla, made a self-portrait by taking a photograph of herself in a mirror. She also made the gesture for "woman" in sign language, pointing to herself, after putting on "makeup" with a piece of chalk.

Tetsuro Matsuzawa, the Chimpanzee Aï, and Her Son Ayumu

At the same time, other researchers gained recognition for the originality of their approach. In Japan, Tetsuro Matsuzawa adopted a complementary approach of research in the field and the laboratory.[36] Striving for a better understanding of human nature and of the origins of behavior and the human spirit within an evolutionary perspective, he conducted studies in comparative cognitive science between humans and apes in Japan and Guinea. Located half an hour outside Kyoto at Inuyama, his Primate Research Institute (created in 1967 by Kinji Imanishi) is home to several colonies of macaques, as well as a group of fourteen chimpanzees which has been followed over three generations. He has known the female chimpanzee Aï, who is at his institute, since she was one year old. Having established a trusting and respectful relationship with her for over thirty years, he first taught her to manipulate visual symbols in order to name objects and colors and to recognize arabic numerals. Aï used computers and read *kanji* (a system of Japanese writing). The research team closely followed the development of Aï's baby, Ayumu, born in 2000. The young male was taught mathematical progressions.[37] He first learned to memorize a sequence of numbers from 1 to 9 on a touch screen. The second stage consisted of presenting him with the same series, each time placed in a different manner on the screen. Ayumu was able to touch the numerals in order from the least to the greatest. Finally, a further difficulty was introduced: some of the numbers were shown only for several tenths of a second before being masked. The short-term memory of the chimpanzee was thus tested as well as his knowledge of mathematical progressions. Humans taking part in the same exercise did not do as well as Ayumu, who was also much faster.

The same team showed that yawning was contagious among chimpanzees. Six individuals were shown videos of other apes yawning. Two of the chimpanzees tested began to yawn as well, a spontaneous reaction interpreted as a sign of empathy. It has been said of Matsuzawa that he "closely observed chimpanzees, Man, and our planet, with an equal devotion."[38] This is an expression of the researcher's evident sympathy and rapport with apes. In Japan, chimpanzees are greeted politely as *o-saru-san* (Mr. Ape) at the beginning of work sessions.

In addition to his laboratory studies, since 1986 Matsuzawa has observed a group of chimpanzees in the Bossou forest in Guinea. This group has been habituated to the presence of scientists over thirty years. These studies focus primarily on the use of tools: "fishing" for ants with a stick, leaves used as receptacles, breaking open nuts with a hammer and stone anvil. The Japanese scientists have shown the use and transport of tools, the diffusion of different techniques through social learning, and the transmission of knowledge, techniques, and traditions among relatives and troop members, known as "learning by apprenticeship with a master." This is not active teaching, with explicit instruction from teacher to student, but learning through observation of the master, the holder of knowledge, the student then trying himself to reproduce the same gestures through imitation.[39]

Frans de Waal

Born in Holland in 1948, Fransicus Bernardus Maria de Waal was interested in animals from an early age. He studied biology at Nijmegen University and there discovered ethology, as well as one of its founding fathers, the Dutch zoologist Nikolaas Tinbergen, who greatly inspired him. From 1975 to 1981, de Waal worked at the chimpanzee colony at the Arnhem Zoo (then one of the world's largest colonies of outdoor-living chimpanzees) under the direction of the Van Hooff brothers. One of them, Anton Van Hooff, was the director of the zoo, and the other, Jan A.R.A.M. Van Hooff, was a professor of ethology at Utrecht University, where de Waal got his doctorate. Marking his association with the Dutch school of ethology, Frans de Waal would dedicate his first book, *Chimpanzee Politics,* to Jan Van Hooff.[40]

In 1981, he left to study rhesus macaques and capuchin monkeys at the Wisconsin National Primate Research Center. At the time, he also made observations of bonobos at the San Diego Zoo and studied the phenomenon of food sharing at the research station of the Yerkes National Primate Research Center in Atlanta (Emory University). He gradually became an authority in the field of observation of captive primates and was even listed as one of *Time* magazine's "100 Most Influential People" in 2007. In 1991 he accepted a position at the Yerkes National Primate Research Center and worked on capuchin monkeys and chimpanzees living in social groups and allowed semi-liberty. A professor in the psychology department of Emory University, he is also the director of the Living Links Center, established in 1997 and associated with the Yerkes Center.

In *Chimpanzee Politics* (1982) and *Peacemaking Among Primates* (1989), he led his readers through a labyrinth of strategies, alliances, and coalitions. He showed, for example, how chimpanzees were capable of acting with Machiavellian intelligence, and he described the phenomenon of reconciliation among great apes. He revealed their capacity for anticipation, cooperation, and trickery, and also demonstrated the importance of females, especially concerning the regulation of relationships within the group and the transmission of sophisticated social skills to their young. Frans de Waal admitted the presence of emotions and sentiments in animals.[41] In *Good Natured: The Origins of Right and Wrong in Humans and Other Animals* (1996), *The Ape and the Sushi Master* (2001), *The Age of Empathy* (2009), and *Are We Smart Enough to Know How Smart Animals Are?* (2016) he did not hesitate to introduce concepts considered specifically human in his descriptions of primates: morality, culture, empathy, and intelligence. Adopting a risky position for a scientist, de Waal brought back into question the principle of parsimony, tied to the work of Morgan and placed at the center of the discipline of ethology.

Humans and Apes

From his first work onward, de Waal has embraced a logic of continuum and resemblance between humans and apes, that fundamental resemblance culminating with bonobos, which are one of his specialties.[42] All his work aims to break down the border between humans and animals. He denounces the overly dualist reading of a world strictly divided between animals and humans, nature and culture, wild and domestic, savage and civilized. Putting forward their common characteristics, he has drawn parallels between the behavior of primates and humans in politics, morality, culture, and sexuality. He insists on the biological roots of human behavior and defends the idea that qualities long thought to be the sole province of humans are in fact products of evolution: culture, empathy, moral conscience, politics, a sense of justice.

However, these notions are difficult to define. They refer to diverse and nuanced realities that vary from species to species. Human beings and human society can neither be reduced to biological roots nor entirely explained through biological methods. Thus, Jacques Vauclair does not deny the evolutionary and diachronic continuity between humans and apes. However, he believes that human cognition, inscribed within a specific history and specific cultures, is tied to unique modalities of communication and representation.[43] Language and its extraordinary power are often invoked as criteria of difference, but Vauclair adds that there are a number of other forms of expression proper to humans, as well as complex achievements very rarely present in the animal world. He thus regrets that certain scientists reduce human specificity to purely biological ends.

Frans de Waal thus showed that all that which was considered the exclusive domain of human culture had too long been cut off from its biological roots. In the human sciences, the dominant viewpoint saw politics, morality, and empathy as strictly cultural, denying them any biological dimension. For de Waal, these were products of evolution. However, the split between humans and animals seemed to most scientists to be insuperable. De Waal's criticisms of the human sciences, whose blind spot is in effect the animal, were in this way legitimate. He believed that certain concepts and themes traditionally, since Galileo, the objects of philosophy and the social sciences (values belong to philosophy, the things of the world to science), should be approached with tools taken from biology.

It was thus nothing less than the grand division between science and philosophy instituted by Galileo that de Waal sought to question. In response to the posture that radicalized an artificially constructed separation between humanity and animality, it seemed that it would be more fruitful to create another axis of reflection, which would, first, call into question an overly strict dualism and, second, take into account both the continuities and the discontinuities between humans and animals — or, even better, between humans and a specific species (rather than the abstract category of "animals," in the sense of what Derrida calls "any living thing that is not man"), all the while making sure not to hierarchize the differences.[44]

New Directions for the Scientific Study of the Great Apes

In a century of research, we have gone from primates that resemble us but are nonetheless said to be fundamentally different to great apes that are extremely close to humans on all levels, from genetic to behavioral, emotional to cognitive. The scientific advances and a change in sensitivity toward animals have positioned them in a radically different manner in the great Western topos. The collective representation of apes has been completely overturned. Furthermore, the idea of great apes' particular personalities has little by little appeared in the world of primatology. Some research has become more centered on individualities, rather than solely general tendencies linked to the species. This orientation toward a singular ethology is nonetheless marginal within studies of animal behavior in the laboratory, which remain for the most part generalist and marked by classical determinations connected with experimental designs.

The question of social intelligence also constitutes a more recent theme. Some researchers have put forth the hypothesis that the "large brains" of primates could have developed and been selected in relation to an extremely sophisticated social world made of competition and cooperation, rather than strictly because of ecological conditions. Their cognitive abilities thus may have above all been fashioned by highly complex socialities.

Studies have shown that the level of intelligence is correlated with the level of social sophistication as a function of the complexity of life in society: in "primate evolution, group life exerted strong selective pressure on the ability to form complex associations, reason by analogy, make transitive inferences, and predict the behavior of fellow group members."[45] Andrew Whiten and Richard Byrne, anthropologists, psychologists, and primatologists at the University of Saint Andrews, audaciously proposed applying the idea of "Machiavellian intelligence" to societies of great apes.[46] The French philosopher Dominique Lestel describes this intelligence, which is both social and political, and heretofore thought of as exclusively human, as the possibility for chimpanzees to manipulate the intelligence of another.

What Is an Ape?

Parallel with a biomedical, neurophysiological, and anatomical orientation of the discipline, a field of study developed that was associated with the naturalist tradition, drawing inspiration notably from Darwin. Adopting the methods of natural history, mainly observation and description, a naturalist "ethology" applied to primates began with Romanes, Garner, and Boutan. Using this type of method, Köhler showed himself to be an extraordinary observer of chimpanzees and orangutans. He saw that which escaped the attention of others. His research was directed by his ties to philosophy, Gestalt psychology, and German studies of nature: he worked with individuals, taking into account the environment in which they lived in a sensory manner, going back to the perspective of von Uexküll and the notion of Umwelt. Like Köhler, Yerkes and Kohts were observers as well as experimenters. They were moved by a genuine curiosity regarding primates and their abilities. Laying the foundations for the discipline of primatology, their work anticipated future research on cognition, social behavior, and the use and manufacture of tools.

Thus we have the outlines of a landscape peopled with various types of researchers. All of them were careful about doing science, but their objectives, their relation to primates, and their methods varied enormously. Opposed to an instrumentalization of the living, and thus to certain modes of experimental research pushed to the limits, some scientists proposed other avenues of research. At the beginning of the 1960s, field research marked an essential turning point in primatology. Primates gradually came to be seen as individuals with their own interests, socialities, and ways of being in the world. Evolutionary psychology addressed specific cognitive skills tied to various specific ecological challenges. During the 1970s, cognitive ethology raised questions of mind, self-awareness, and the guesser-and-knower paradigm of intersubjectivity, which became objects of scientific research. In laboratories, new experimental devices emerged, particularly spaces where primates lived in social groups and had a degree of liberty. Researchers began to address questions that aroused the interests of the primates themselves, and made more sense for them. They had the opportunity, in these places, to take initiative, show their preferences, call on their abilities and, sometimes, form relationships with the humans studying them.

Some researchers, philosophers, and scientists (particularly primatologists) also questioned the prohibition of anthropomorphism by delving into what this concept actually covers, without ceding any scientific rigor. The German philosopher and zoologist Adolph Portmann advocated a *critical anthropomorphism* tied to a form of assumed proximity, of empathy, toward the living (*Einfühlung*), previously proposed by Husserl. Later taken up by Japanese researchers, this anthropomorphism constitutes, according to Portmann, a privileged mode of access to the animal. Concerning great apes, Frans de Waal argues that, given the close relation of humans to apes, anthropomorphism is inevitable.[47] The sequencing of the genomes of different species of apes confirmed this close proximity genetically. For de Waal, it was thus declaring apes as being fundamentally different that required justification, rather than the reverse.

Little by little primatologists challenged the idea that certain capacities belong only to humans: communication and aptitude for symbolic language, different strategies (Machiavellian power politics, conflict resolution, reconciliation behaviors, alliances, organized hunts), the guesser-and-knower paradigm of intersubjectivity, empathy and moral tendencies, and stages of life characterized by, for example, a midlife crisis and menopause. As field research became more widespread, our knowledge of primates grew, with evidence accumulating of sophisticated social organizations (with social reciprocity, system of corruption, adoptions, territorial defense, behaviors of sharing, and so on), means of specific transmission (social learning), cultures, interspecies contacts, and tool technologies. By the beginning of the twenty-first century, apes had finally earned the right no longer to serve as guinea pigs.

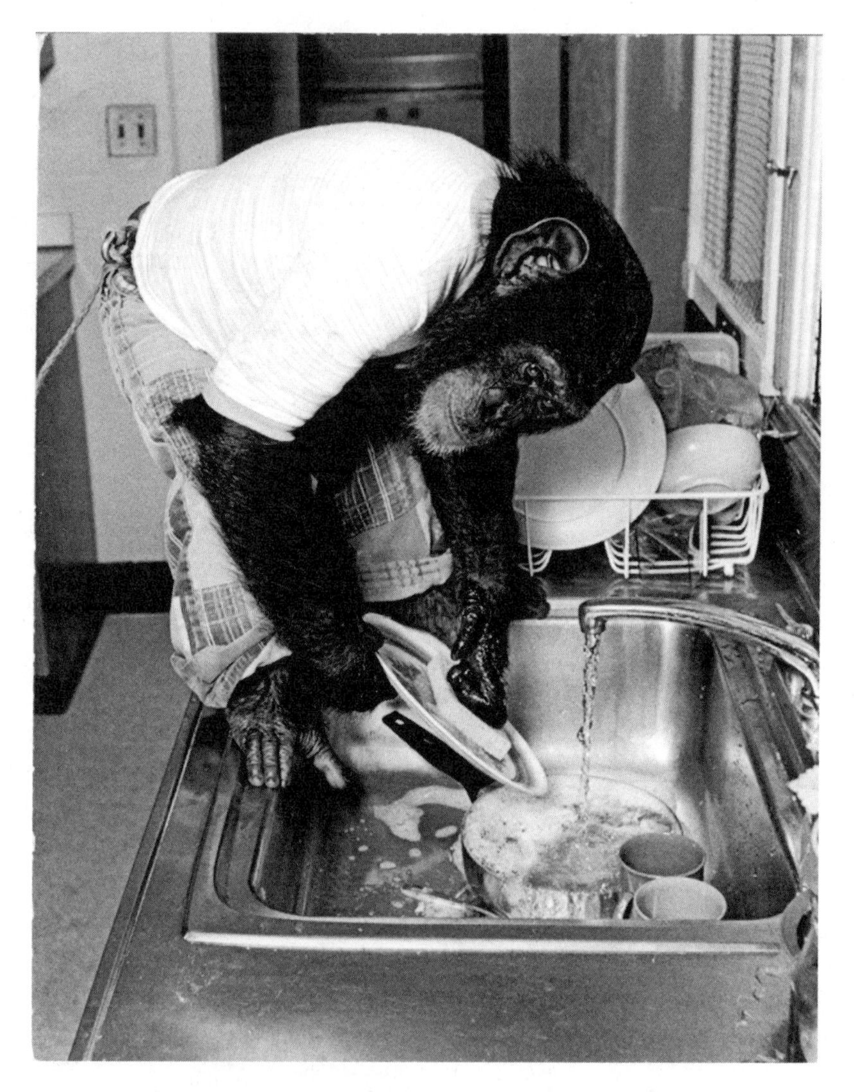

Nim Chimpsky washing dishes.
(From Herbert Terrace, *Nim* [1979]. Photograph © Herbert Terrace)

FIVE

Apes That Think They Are Human

Astronaut Apes, Painting Apes, Talking Apes

AT THE TURN OF THE TWENTIETH CENTURY, encounters between humans and great apes became more common, whether at zoos, within human families, or in research facilities. Primates began to be better known to the general public, particularly chimpanzees, which elicited somewhat more sympathy than gorillas or orangutans. In Cincinnati between 1888 and 1894 the chimpanzees Mr. and Mrs. Rooney were shown to visitors dressed in their Sunday best. In the 1920s, the Bronx Zoo exhibited its apes having dinner, while Detroit featured a primate "happy hour." At the London Zoo, primate "teatime" was famous. The chimpanzees used spoons, cups, and teapots, to the great joy of the public. Human dignity, closely linked to the maintenance of separation between humanity and animality, was saved because the apes finished by lamentably failing to imitate humans: once the guardians, figures of human authority, turned their backs, the chimpanzees spilled the tea, broke dishes, or made fools of themselves by using certain accessories inappropriately. In reality, their prowess was even more remarkable given that they were perfectly able to use all these objects: they were only pretending to be maladroit, conspiring with their handlers to manipulate the crowd.

In this chapter we will see how humans and apes have created very close and subtle ties in certain places far from biomedical laboratories or the sites of experimentation discussed in previous chapters. Inside these communities, relationships take on different forms: slices of life shared between keepers and apes in zoos, the adoption of primates by individuals, families, or researchers in a shared space. These different ways of living together all lead to apes' adoption of human ways of being and living. The places of coexistence between humans and apes have also made visible an essential common world.

Apes That Imitate Humans

At the end of the nineteenth century, Dr. Maclaud, a medical doctor in the French Navy, was sent to Guinea. He described the behavior of the male and female chimpanzees Baboun and M'Balou. Baboun had learned how to sew and dance. He was able to sign and seal documents. He also knew how to use matches. M'Balou, the female, was given different household chores. This story, along with numerous other descriptions of primates taking on human behavior, was related by Henri Coupin in his book *Singes et singeries* (1907). A doctor of science, Coupin made certain that his documentation was rigorously exact. He told, for example, of a female chimpanzee aboard a boat that was capable of filling all the roles of a sailor. In the kitchen, she lit and watched over the stove, knowing how to evaluate the temperature, and communicate this to the cook, who had perfect confidence in her. Consul, a chimpanzee at the Manchester Zoo, wiped his mouth and washed his hands after meals, put on a hat to go out, put coal in the stove, and smoked a pipe and cigars. Coupin also described the exploits of a chimpanzee shown in a Parisian theater around 1903. Dressed in an English suit, the chimpanzee greeted the spectators by tipping his hat. He walked on two feet, smoked cigarettes or cigars (striking the match himself), rode a bicycle, drove an automobile, and used a typewriter. Invited to a reception, he shared a table with other guests.

On the other side of the Atlantic, various authors reported cases of primates that were excellent in imitating human behavior. In 1907, the chimpanzee Consul II was invited to a gala dinner among the polite society of Jacksonville. The star of the evening, he drank champagne and behaved perfectly at table. He showed such intelligence that certain people even thought about operating upon him to allow him to talk. William Hornaday, the director of the New York Zoological Park (today the Bronx Zoo), mentioned the exploits of a six-year-old chimpanzee seen in a New York theater in 1909. Peter was capable of stringing together more than fifty actions: he

The female chimpanzee M'Balou sweeping the floor.
(From Henri Coupin, *Singes et singeries,* 1907)

smoked, brushed his hair, powdered his face, roller-skated, and used different tools with great dexterity.

Consul and John Daniel

At the beginning of the twentieth century, Westerners delighted in these imitative behaviors and tried various ways of staging them, including at the

Ménagerie at the Jardin des Plantes in Paris. In 1903, a young chimpanzee named Consul was exhibited there as an ape savant. His trainer, Franck Charles Bostock, taught him how to sit at table, clean the floor, get dressed and undressed by himself, and put away his hat and clothes in the dresser. Consul could put on and take off his shoes, then put on slippers. He smoked a cigar, rode a bicycle, and drove a car.

Sometimes people took great apes into their homes. Tied to a strong interest in primates, their motivation was often spurred on by a desire better to understand the process of hominization as well as to know to what point the intelligence of these animals, among the highest evolved, could be developed within a stimulating human environment. In 1909, Richard L. Garner was among the first to bring apes into his own home: two chimpanzees in Liverpool, and later several chimpanzees and a gorilla at his home in New York. A short time later, a young western lowland gorilla arrived in London. Coming from French Gabon to France in 1917, he was bought by the trader Hamlyn, who gave him his own first names: John Daniel. Hamlyn also had chimpanzees that he dressed and let sit at table. The gorilla was then bought by the store Derry and Toms to entertain crowds at Christmas. Major Rupert Penny of the Royal Air Force and his aunt Alyce Cunningham were fascinated. They decided to adopt John Daniel and bring him into their home.[1] John Daniel would live with them from December 1918 until March 1921. It is without doubt the first case of humans and gorillas living together in Europe.

Alyce Cunningham's goal was to raise Johnny Gorilla as a child and make him an "ordinary member of the household." About two years old, John Daniel had his own room with electric heating in an apartment on fashionable Sloane Street. He earned his nickname of "the civilized gorilla."[2] He learned how to sit correctly at table, wash himself in the bath with soap, comb his hair, use a toilet, and put away his things. Once an "apprenticeship" of six months was over, he was allowed to move about freely. He got into the habit of welcoming visitors, taking tea with them, and going to children's parties. He loved to play hide-and-go-seek and blindman's bluff with Alyce Cunningham's three-year-old niece. He also loved to jump on the bed. Sometimes he was taken to the London Zoo. When he accompanied Cunningham to the family country home in Gloucestershire, John Daniel rode in the train with the other passengers, without a leash or restraints.

Eventually, however, he became a heavy burden to bear because of his refusal to be separated from his adoptive parents, even for short periods of time. He was therefore sold and sent to the United States. The person who escorted him, Mr. Benson, knew nothing of primates. Badly taken care of and lonely in Madison Square Garden, John Daniel fell ill and died in April 1921. He was dissected by scientists from the American Museum of Natural History in New York. His skeleton was conserved in the museum's collections and his skin exhibited as a natural specimen in the Hall of Primates.

In 1923, Alyce Cunningham acquired another gorilla, Sultan, which she called John Daniel II. The following year, she took him on tour in the United States and Europe in collaboration with the Ringling Bros. and Barnum & Bailey Circus. This time, Cunningham never left the gorilla: they traveled together, whether by train, car, or boat. They shared the same hotel room and visited the American Museum of Natural History together, where John Daniel II found himself face-to-face with the exhibit of his preserved predecessor. He, however, also died young, in 1927.

The Ape of the American Museum of Natural History, Meshie Mungkut

Another famous case of a great ape being adopted by a human was the female chimpanzee Meshie Mungkut, brought back from Cameroon by Henry Cushier Raven, one of the greatest collectors of the time and associate curator at the American Museum of Natural History. Meshie became famous through a film, articles, conferences, and a tour of various cities in the United States.[3] In Raven's film, Meshie is shown in the master bedroom on the bed with Raven's children, Jane and Henry. Another scene has them sitting at a table in the garden sharing a glass of fruit juice. During the winter, she was filmed pulling a sled on the snow, walking upright. Several scenes in the film show Meshie's talent for imitating human behavior. Anxious to do well, she took great care, for example, in feeding the Ravens' newborn

Henry Raven and Meshie. "I first heard her called Meshie Mungkut. I was told that they had given her that name, which in their language was a nickname for a chimpanzee and was especially applicable to a small one that fluffed up its hair so that it looked big, though in reality it was very small. The name Meshie seemed a good one and has been hers ever since."
(From Raven, "Meshie," 1932; photograph courtesy of Henry M. Raven)

daughter Mary. Skillfully riding her tricycle and keeping watch over her shoulder of the others' progress, she led a parade of a number of the neighborhood children on their bicycles.[4]

However, although Meshie spent nearly four years in the family home, from February 1931 to December 1934, she was not really a member of the family. According to the curator's son, Henry Mercelis Raven, the filmed scenes were, for the most part, staged to endear Meshie to the general public—and to advertise the museum. In reality, she was kept in a cage in the basement of the family home in Baldwin, Long Island. She was let out only when Raven was present because she wouldn't obey anyone else.

Meshie sometimes accompanied Henry Raven to the museum. The visitors were astonished, and certainly charmed, to see a chimpanzee riding a tricycle through the exhibit halls. At noon, she ate at the curator's table, conducting herself well. She was even invited to several gala dinners, one of them given by the president of the museum at the Waldorf Astoria Hotel. She behaved perfectly throughout the reception.

Unfortunately for Meshie, the person she undoubtedly considered as her adoptive father (she spent almost a year alone with him in Africa) had to leave for many months on a trip to Burma (today known as Myanmar). During his absence, his wife, Yvonne Aurousseau, refused to take care of this rather peculiar fourth "child," particularly one so unruly and at times even dangerous. Raven sought a place for Meshie, who was finally admitted to the Brookfield Zoo in Chicago. She died there in 1937 at the age of eight, due to complications in the birth of her daughter, Eve. The public can see her mounted specimen on display in the Hall of Primates at the American Museum in New York, next to that of the gorilla John Daniel I.

During the colonial period, expatriates also adopted primates and raised them as their own children. For instance, in the former Belgian Congo, Madame Trompet took care of Malenga. However, upon her return to Belgium she was forced to put her in the Antwerp Zoo. Having been admitted as a chimpanzee, she died there as a pygmy chimpanzee (or bonobo), this new species of great ape having been described by Ernst Schwarz in 1929 (as a subspecies) shortly after her admission to the zoo.

The Colony of Rosalía Abreu

Abreu was a true pioneer among the very restricted circle of those who created private primate colonies. She began taking in primates toward the end of the nineteenth century. While she loved all animals, she had a marked preference for the great apes. In thirty years, she would take in over fifty chimpanzees and orangutans, as well as dozens of monkeys, at her house Las Delicias in Quinta Palatino, on the outskirts of Havana. At a time when the life expectancy of captive primates was very low, this was the first time that representatives of these species attained a longevity comparable to that they achieved in their natural environment. Living in a manor house with an immense forested park, Abreu provided enlightened care for her residents and perfected various norms for dietary requirements, hygiene, cage materials, and social environment. The primates lived in some fifty large, well-ventilated cages, protected from drafts, that had both shaded and sunny areas, with a basin for a toilet as well as a room with beds and mattresses for the great apes. They had abundant spring water as well as a varied diet of vegetables and fruit, and what was left uneaten was taken away.

Abreu also understood that primates were social beings: she made sure to put them in cages with companions, either of their own species or of another. She gave them activities that allowed them contact with humans. Some of the apes slept in her house (sometimes in her room) and ate at the table. She made sure, moreover, that they had "good table manners." Robert M. Yerkes initiated a correspondence in July 1915. Abreu gave him a long response. She described the birth of Anumá, the first chimpanzee to be born in captivity, on April 27, 1915, and added photos of the baby in the arms of her mother, Cucusa, captured in Sierra Leone (Anumá's father, named Jimmy, had been bought in London).

Yerkes and Abreu maintained a correspondence until her death. The first observations of the behavior, growth, reproduction, birth, and social life of different species of primates were conducted in the colony. Yerkes visited Cuba several times. A large part of the sources for the book *Almost Human,* written by Yerkes and published in 1925, derive from these observations. Thanks to Abreu, the care given to captive primates became much better and their life expectancy increased. Upon Abreu's death in 1930, the colony was gradually broken up by her son Pierre, who kept up an exchange of letters with Yerkes for many years. Many of the primates went on to Yerkes's laboratory.

Gertrude Lintz and Gargantua the Great

Gertrude Ada Davies Lintz also fell in love with the great apes. At the start of the 1930s she had the largest collection of apes in the United States at her house on two acres of land in Brooklyn. She treated them like they were her own children. She taught them to eat correctly at table and bought their clothes and shoes at the most elegant shops in New York. Lintz was not, however, simply looking to fill an empty nest. She was genuinely captivated by apes. She undertook one of the first experiments regarding language learning with the chimpanzee Susan and raised the possibility of better understanding human evolution through the observation of primates.

In 1931, a friend of the family, Captain Arthur Phillips, offered Lintz six chimpanzees and a gorilla. Upon his arrival, the young gorilla Massa was suffering from pneumonia, but Lintz and her husband, a doctor, cured him. In 1932 she took in another gorilla, also in very bad health and almost blind. Buddy, born in the Congo around 1930, had been given as a present to Captain Phillips and became the mascot of the ship. A rictus deformed his mouth, which made his teeth show and gave him a terrifying appearance. Legend has it that he was hit by lightning. In fact, a sailor, seeking vengeance after being sacked, threw acid in Buddy's face, partially disfiguring him. Even though he had a surgical procedure to repair the damage, Buddy kept the scars of this aggression all his life.

In 1935, Lintz resolved to separate herself from the gorillas in spite of her love for them. Having been badly bitted by Massa, who was now an adolescent, she feared other attacks. At the time, there were only a dozen or so gorillas in the United States. Wanting a companion for Bamboo (the first gorilla to have lived for a long time in captivity), the Philadelphia Zoo evinced some interest. In December 1935, Lintz sold Massa to the zoo for $6,000. Both parties to the transaction were under the mistaken impression that he was a female. Nevertheless, the gorilla stayed in Philadelphia until he died in 1984 at the age of fifty-four. He was at the time the oldest gorilla in captivity in the United States.

Lintz later offered Buddy to the Ringling Bros. and Barnum & Bailey Circus. Sold for around $10,000, he joined the circus in 1937, where he received the impressive name of Gargantua the Great in reference to the work of Rabelais. The peace-loving Buddy was shown as "the most powerful monster" ever captured by humans and "the most terrifying creature in the world."[5] He fit his role perfectly, his scars giving him a fearsome appearance. Gargantua became the star of Ringling Bros., achieving great success that helped the circus to survive during the Great Depression. In 1941, a "marriage" was arranged between Buddy and Mademoiselle Toto, which was called the "future Mrs. Gargantua" or "Gargantua's fiancée." Mr. and Mrs. Gargantua never mated. Toto refused even to share the same cage as her betrothed. The skeleton of Gargantua, who died in 1949, has been part of the collection of the Yale Peabody Museum of Natural History since 1950.

A Gorilla in Cuba

The circus's press release made the future Mrs. Garagantua the most well-known female gorilla in the country. But who was Toto? Her history began with the killing of her father by Kenneth and Maria Hoyt during a safari in French Equatorial Africa in 1931.[6] Sent to collect the skeleton and skin of a gorilla for the American Museum of Natural History, the couple was among the few Westerners at the time to have observed this species in the wild, and to have known the infant ape they adopted while she was yet in her native land. They decided to settle in Havana. Maria Hoyt took care of Toto (*m'toto* means "baby" or "small child" in Kiswahili) as if she were her own child. She built her a house in the garden of their sumptuous home. Baby Toto had her own bedroom, playroom, and enclosed yard. She was very affectionate, behaved well at table, rode a bicycle, combed and washed herself, brushed her teeth, wore clothes, and used a toilet. She also loved to dance. She even adopted two cats, Blanquita (which followed her everywhere) and her son, Principe. The gorilla also liked to play with one of the Hoyts' bull terriers. Toto was taken care of by one of the former keepers of Rosalía Abreu, José Thomas, who slept in Toto's room, fed her, walked her, and played peekaboo and hide-and-go-seek with her.

Weighing only nine pounds upon her arrival in Cuba, Toto would reach 438 pounds at around age eight. Her actions could sometimes be aggressive. The household lived more and more under the domination of the "hairy little black child," as they called her. Kenneth Hoyt died in June 1938. With a heavy heart but fearing an irreparable accident, Maria Hoyt was forced to accept an offer from the Ringling Circus. Toto was sent to Florida. After having followed the circus from town to town for six months, Maria Hoyt returned to Cuba. Toto, who did not take part in any of the circus shows, was placed in a house in Sarasota, where she kept her human habits. She died in 1968 of a brain tumor in Venice, Florida.

The Noells and the Gorilla Show

Mae and Bob Noell bought their first chimpanzee, Snookie, in 1940.[7] They thought of their gorillas, chimpanzees, and orangutans as real members of their family. Their chimpanzee Joe could ride a bicycle, tricycle, and scooter. He could also walk on his hands. Mae Noell particularly loved Topsy, a baby gorilla that went everywhere with her. She cradled Topsy in her arms, danced the waltz with her, and described her as "my most precious baby."

Their charges were also employed in a "show of athletic apes." The apes boxed with a number of spectators who wanted to challenge them. The goal was to last more than five minutes against the apes, who were taught at a young age the basics of the sport by playing games with boys of the same age. In order to protect the human participants, the Noells gave their apes gloves, leather masks, and shoes. Apart from several scratches, none of the visitors were hurt. According to Mae Noell, the boxing apes were never beaten. The Noells were among the pioneers of raising orangutans in the United States. They had even more reproductive success than the zoos. During the 1960s, their private collection of thirty apes was the largest in the world. Forty-five young were born between 1948 and 1978, among them the chimpanzee Lucy, who could communicate using language, in 1964.

The Golden Age of Painting Apes

Little by little, humans and apes were thus led to live in common spaces and find modalities of cohabitation. This living together would have a remarkable consequence: the primates showed an extraordinary ability to appropriate certain human habits and savoir-faire. In part tied to phylogenetic proximity, this ability would be examined by scientists outside of traditional laboratories, in studies marked by a liberty and creativity that would rarely be equaled.

The painter and primatologist Desmond Morris, author of the best seller *The Naked Ape,* was interested in the appeal of drawing and painting for primates. During the 1950s, considered the "golden age of painting by apes," he counted about thirty apes that painted in the United States, England, the Netherlands, and Germany. In addition, each primate had a personal, easily recognizable style. For example, the author and painter Lilo Hess, adoptive mother of the chimpanzee Christine, described her work as "quite beautiful in their original colors." Hess's photographs show Christine finger painting, making use of a rainbow of colors. It was one of her favorite hobbies. From 1956 to 1962, the young chimpanzee Congo, one of the most famous of the ape painters, made some four hundred paintings and sketches. All of this hard work was evidently decisive in the process of an emerging expertise. Over time, Congo was able to concentrate on his painting for over an hour. His manual precision and his sense of composition grew more assured and refined.[8]

The instructor of the "talking" ape Washoe, Roger Fouts, said that chimpanzees love to make art. Belonging to the group of chimpanzees raised by Fouts, Moja was known as the first ape to paint figurative works: after having traced a form on a page, she signed the word "bird." She also drew a circle, colored in orange, and made the sign for cherry. The female gorilla Koko made a painting that she said, by signing, was a bird. Fouts also said that his chimpanzees gave their paintings titles. Washoe, for example, invented the phrase "Electric Hot Red," red being her favorite color.

When Apes Began to Speak

During the 1960s, American researchers began a number of language-learning programs for great apes.[9] These projects were preceded by two studies that allowed the testing of a new experimental model: the integration of a young ape into a human family as a full member, and the education of this infant by human parents (cross-fostering). It is thus a process that consists of raising the young of a given species by adults belonging to another species. The first experiment was conducted with the chimpanzee

Gua (1931–32), and the second with a young female of the same species, Viki (1947–54), although this time with the goal of teaching her to "speak." The inability of Viki to pronounce more than four words after seven years was undoubtedly one of the reasons this type of experiment was suspended for more than fifteen years.

During this hiatus, scientists finally took into account what some scholars had long understood: apes do not have the possibility of *speaking*, mainly because of a larynx positioned differently from that of humans and the significantly lesser flexibility of the tongue. In the seventeenth century, Pepys had in fact already advocated teaching sign language to chimpanzees, as did La Mettrie and Monboddo in the eighteenth and the psychologist Wilhelm Wundt in the beginning of the twentieth. It would not be until the start of the 1960s, however, that a number of American researchers initiated various scientific projects involving learning among the great apes. Human language was taught to apes in the form of sign language, symbols (accessible mostly on computer screens), or words made of colored plastic.

The instructors again had the idea to integrate the primates into human families because they believed, with good reason, that this aspect was essential to the acquisition of language by our closest cousins. Within these different studies humans and apes would live together, forming real communities within the framework of official experiments. The apes were positioned not only to communicate with their instructors and appropriate certain elements of language but also to adapt perfectly to their environment and adopt certain human behaviors to a degree few would have believed possible. However, obsessed by the question of language and the theoretical debates surrounding language learning by apes, the researchers neglected this formidable life experience. All their attention was focused on the scientific and performative aspects of these projects (the human species was finally able to make animals talk, allowing them to communicate with another species), leaving in the shadows this essential moment in the history of relations between humans and animals.

The Chimpanzees Gua and Viki

Gua was born on November 15, 1930, in Cuba at the colony of Rosalía Abreu. Shortly afterward, she was sent to Yerkes's Anthropoid Experiment Station (Orange Park, Florida) together with her parents (Jack and Pati). Although before her the female Alpha had been the first primate raised by humans *in a laboratory* (that of Yerkes), Gua was the first great ape officially brought into a human family as an experiment. She was educated from seven and a half months on as a human baby in the home of Luella and Winthrop Niles Kellogg, scientists at Indiana University, until March 28, 1932, when the experiment was ended. What is more, she grew up with a child, their son Donald, aged ten months when she joined the family.

The two babies were fascinated with each other from the outset and loved playing together. The Kelloggs compared both their behavior and their development. In this regard, the book published in 1933 by the Kelloggs is extremely precise, well documented, and rich in reflections upon the experiment under way.[10] After nine months, the young ape and the infant began to imitate one another. The child's imitation of the chimpanzee reached such a level that Donald was late in speaking. The Kelloggs stopped the experiment after nine months. Gua returned to the laboratory directed by Yerkes, where she died of pneumonia at the age of three, on December 21, 1933.

On September 1, 1947, Keith J. Hayes and Catherine Hayes, two American psychologists from the same laboratory, took in the chimpanzee Viki a few days after her birth.[11] They took care of her for seven years. They wanted to test her intelligence and teach her how to speak. They did not at first realize that apes are incapable of speech for anatomical reasons, which they discovered during the experiment. At the end of her stay with the researchers, Viki was able to pronounce only four words: "Mama," "Papa," "up," and "cup." The Hayeses were greatly disappointed. However, Viki spontaneously used cards with images to let her human companions know what she wanted. She showed Cathy Hayes a cup and an ad for drinks, for example, to mean that she wanted to drink a soda. In this manner she herself showed the direction the Hayeses should have taken, to allow her to communicate with them via a symbolic language rather than spoken words. Unfortunately, Viki died of viral encephalitis on May 11, 1954, bringing this fascinating experiment to an end.

In the 1960s William Lemmon of the University of Oklahoma developed the idea of *cross-fostering* based on these two experiments and began a number of projects of adopting young apes into families. Starting in 1962, he himself raised the chimpanzees Pan and Wendy, each one year old, with his own children, ages ten and eleven. He then placed several apes with other American families, beginning in 1965 with Mae and Lucy. Unfortunately, according to Fouts, who worked alongside him for several years, Lemmon had an almost godlike conception of his relation to primates. In order to establish his ascendancy over them, he often resorted to cruel treatment. Pan, who was terrified of Lemmon, paid the price.

Lucy

Lucy, a chimpanzee, was born in 1964 at the home of Bob and Mae Noell. Taken from her mother at only two days old, she went to the University of Oklahoma, Norman, and was subsequently placed with the Temerlin family by Lemmon. Jane Temerlin was Lemmon's secretary, and her husband, Maurice, was a psychotherapist. Lucy, whom Maurice presented as his daughter ("My daughter, Lucy, is a chimpanzee") was raised with the Temerlins' son, Steve, who became like a brother to her.[12] She slept for several months in a crib in the bedroom of her adoptive parents; later she shared their bed, slipping in between them each night. She ate in a high chair and had her own room and toys. The Temerlins gave her a kitten that she thought of as her pet, as she did Namuq, the family dog. She learned to eat correctly, peruse magazines, watch television, open the door for visitors, appreciate wine, and later share cocktails with Maurice. She made tea for visitors: she put water in the kettle, placed it on the stove, took it off when the water boiled, poured the water into the teapot, added the teabag, and served it in teacups. Lucy very quickly learned how to use everyday objects: brush, pencil, mirror, vacuum cleaner, garbage bin, tools, matches, and lighter. Maurice Temerlin said there was no difference between Lucy and a human child. For him, she was simply a little girl who was a bit stronger, and less predictable, than most others. The boundary between humans and apes was blurred.

Lucy learned American Sign Language (ASL) with Roger Fouts, who worked at Lemmon's Institute of Primate Studies, founded in 1965. She knew more than a hundred words and made up formulations for objects whose name she did not know—watermelon, for example, which she called "sugar fruit." She appeared in the pages of several different newspapers and magazines (*Life, Psychology Today,* the *Los Angeles Times,* and the *New York Times,* among others). As an adolescent, she became interested in men rather than other apes. To Temerlin's great surprise, she also used the tip of the vacuum cleaner (which was already turned on) to masturbate.

After more than twelve years of living together, the Temerlins were forced to look for a new home for Lucy. Refusing to sell her to a biomedical laboratory, the fate of many of Lemmon's other primates, they sought another solution and decided to give her a chance to return to her "natural" environment. In September 1977 they accompanied her to Africa, and then returned to the United States. Janis Carter stayed on to help Lucy face the challenges that awaited her. For Lucy, being taken from her human family was undoubtedly a real cultural and affective shock. After her transfer to the Chimpanzee Rehabilitation Trust in Gambia, Lucy continued to ask, by sign, for the things she had known since her childhood: toys, magazines, television, American food. She refused to mingle with other chimpanzees. After ten years, she apparently began to adapt to life as a chimpanzee, although she refused to mate with the other apes. She died in 1987, at the age of twenty.

Washoe

Born in West Africa in 1965, and dying at the Chimpanzee and Human Communication Institute (Central Washington University) in 2007, the female chimpanzee Washoe was one of the most remarkable personalities among the rather small community of talking apes. Captured in Africa, she was transferred to the Holloman air force base in Alamogordo, New Mexico, as part of NASA's space program. She was given the name Kathy. Only two of the chimpanzees in the program, Ham and Enos, were ever sent into space. The other primates were later sent to medical research laboratories.

However, just before Kathy was sent to a laboratory in 1966, Allen and Beatrice Gardner, looking for a young primate to use in a language-learning experiment, bought her, took her to the University of Nevada, and renamed her Washoe. She lived in a trailer in the middle of a large yard. However, human companions were constantly in contact with her. By placing her in a community sharing the same language (ASL) and offering her an interesting environment, the Gardners hoped to be able to have real conversations with her, rather than a series of various requests and limited responses, as was often the case in a stricter experimental framework.

In 1969 the Gardners' first article, published in *Science,* had a tremendous impact: humans had finally realized one of their oldest dreams, to communicate with another species, breaking in some sense their essential solitude. Washoe made the acquaintance of Roger Fouts, a young assistant in the psychology department, in September 1967. This meeting would entirely change both their lives. In 1975 Roger Fouts took over the studies begun by the Gardners. After several weeks he stopped asking himself whether he held a human or an ape baby in his arms. The distinction seemed to him totally without sense.

Washoe quickly learned a number of signs (she knew 132 by the age of four), progressing to combinations of two signs, then sequences of three signs. When she did not know the name of an object, she invented a formulation. For example, when she saw a swan for the first time and was asked what kind of animal it was, she responded, "water bird." She was also able to generalize, categorize, and establish relationships between words. Fouts and Washoe began, little by little, to have real exchanges in ASL. In the preface to the book *Next of Kin,* Jane Goodall described her fascination as she observed the communications between Washoe and Fouts. It was an actual exchange, a real sense of communication, reinforced no doubt by its symbolic nature. Moreover, in Washoe's group, language use was spontaneous, including between the primates themselves.[13]

As an adult, Washoe taught some elements of sign language to her adopted son Loulis without any human intervention. What is more, she did it using the same method the researchers had used (called molding): she demonstrated and repeated the gesture, then corrected her son's gesture by repositioning his fingers. This transmission was all the more remarkable in that it is extremely rare among primates, even apes. The young assimilate the necessary knowledge to live in their environment by a process of saturation, observation, social emulation, and imitation rather than by a teacher-student method such as Washoe used with Loulis, wherein a master transmits her knowledge to students in an "active" manner.

Sarah, Koko, and Nim Chimpsky

Other research was taking place at the same time. The American psychologist David Premack directed a program using nine chimpanzees, among which five were taught to communicate using a series of colored plastic tokens. Born in Africa in 1962, the chimpanzee Sarah took part in these programs. She lived surrounded by the researchers and was close to Premack's wife Ann, although she did not have the simulation of a real family life.[14] Undertaken at the University of California at Santa Barbara, this study was based on an experimental paradigm and thus did not facilitate the same type of exchanges as in the case of talking apes raised by humans. In this school-type setting for learning, the emphasis was placed on word order in phrases, which is essential for comprehension. The chimpanzee was also capable of making comparisons (for example, using the comparative "same") and composing either negative phrases ("No . . .") or conditional phrases ("If . . . then . . ."). Some apes in the groups, Gussie or Walnut, for example, never learned any words at all.

Born in 1971 at the San Francisco Zoo, Koko was the first gorilla able to communicate in human symbolic language. Francine Patterson was working at the zoo as a student. Having been given permission to take Koko home every night, she began to teach her sign language in 1972.[15] According to Patterson, Koko knew approximately a thousand signs (Gorilla Sign Language) and two thousand terms in English. She passed the mirror self-recognition test, which most gorillas failed. She was among those rare animals that had a pet, in her case a cat.

At the same time, the chimpanzee Nim Chimpsky, born in 1973, was placed in the family of Stephanie Lee by Herbert Terrace of Columbia University.[16] From the age of two weeks old, he became a full member of this family with seven children. However, the young ape stayed in the family barely nine months. Substitute parents and instructors followed, and young Nim was undoubtedly troubled by this unstable environment. His results were mediocre. Later put in Lemmon's Institute of Primate Studies, he was then transferred to a biomedical research laboratory in New York (LEMSIP). He died at the age of twenty-six of a heart attack at Cleveland Amory Black Beauty Ranch in Texas, where he had at last been admitted after passing through the hell of the biomedical laboratories.

Talking Apes: Lyn and Chantek, Sue and Kanzi

The talking orangutan Chantek (born in Atlanta in December 1977) was raised by a former student of Fouts, the anthropologist Lyn Miles, beginning in 1978. She formed a real family cell with Chantek. She slept with him and raised him as her own child until the age of nine in order to create a sociocultural framework favorable to language learning.[17] Among the talking apes, the bonobo Kanzi, born in 1980, proved to be exceptional.[18] Kept at the Language Research Center in Atlanta, he also learned to understand and manipulate symbols (called "lexigrams") in an indirect manner, through lessons given to his adoptive mother, Matata. His learning was thus done entirely spontaneously, through saturation, social emulation, and imitation. This research program had been started in 1971 by Duane Rumbaugh, notably with the female chimpanzee Lana (the Lana Project). A number of apes participated in this study, including Sherman, Austin, Panzee, and Panbanisha. When Kanzi was eighteen months old, his instructor, Sue Savage-Rumbaugh, and her sister Liz adopted him and took care of him full-time, so that he would not be traumatized by the loss of his mother. Savage-Rumbaugh slept with Kanzi, as Lyn Miles had done with Chantek. The bonobo communicated with researchers using lexigrams. He was able to use around three thousand combinations of words. As others had done, he evoked absent objects and invented new formulas to describe elements whose name he did not know. He also showed that he had a certain notion of time and seemed to understand another's point of view.

While they evidently have not become experts in grammar, great apes have made significant strides in employing our symbolic languages, even if they use them only partially. Here, as in other domains, apes communicate according to their own modalities, shaped by their interests as well as physical and cognitive capabilities. After all, it is not *their* means of communication. They are nevertheless able to communicate their feelings, desires, perceptions, and questions. Humans, on the other hand, are rarely able to communicate with apes in the apes' own means of expression. These experiments have also made possible exceptional encounters that, in the best of cases, lasted a lifetime. Roger Fouts remained faithful to Washoe until her death. He and his wife Deborah always acted ethically and in the interest of the chimpanzees they had under their protection, dedicating their entire careers to them.

Although these programs were popular, increasing criticism contributed to their death knell. Their high cost, the lack of interest in questions of language, and the problems arising once the primates reached puberty all certainly had an effect on curtailing the research.

The Becoming-Human of Talking Apes

Placed in these family structures, the talking apes showed an extraordinary capacity for adaptation to human environments, appropriating human savoir-faire, customs, socialities, and even hints of a human ethos. All of them used silverware, glasses, and napkins, and all generally behaved very well at the table. Raised by Catherine and Keith Hayes beginning in 1947, the female chimpanzee Viki started at the age of sixteen months to imitate Catherine's daily actions: she dusted the furniture, washed the dishes, and vacuumed. These imitative behaviors grew stronger with age. The young primate blow-dried her hair, painted her fingernails, and used tweezers.

She was very adept at using a pencil sharpener, a bottle opener, and an emery board. Washoe gave her doll a bath, as her instructors did for her. She filled a small basin with water, put the doll in, washed it with soap, and then dried it with a towel. The young female was also able to concentrate for long periods of time on sewing.

In Washoe's group, Moja liked to wear clothes and brush her "hair." The chimpanzee Nim Chimpsky slept next to his favorite doll and sucked his thumb like a human child. He also voluntarily helped his instructors prepare meals, wash the dishes, and put clothes in the washing machine. When his human companions took off his pants to teach him how to use the toilet, he showed modesty by covering his genitals with his hands. For evidently practical reasons, the talking apes learned how to use toilets. The chimpanzee Nim signed the word "Finished!" when he was done. The orangutan Chantek panicked when he no longer had a toilet in his new place of captivity.

Writing and Diving

Certain behaviors of the talking apes demonstrate, moreover, a form of symbolic imagination. Fouts related that one of Washoe's preferred games was to put her favorite dolls in a circle and talk to them using sign language. She stopped as soon as one of the instructors appeared. At the Language Research Center, Austin liked to simulate the presence of food. Sherman "made believe" that his doll King Kong bit his feet and hands. Koko had the habit of reading illustrated children's books "aloud" to herself using sign language. Kanzi spoke to himself using an interposed panel of icons without any human assistance.

The female chimpanzee Panzee, raised with the bonobo Kanzi, carefully traced signs, very close to human writing, on a sheet of paper, taking care to stay within the lines and not skip any. Surrounded from a young age by scientists who observed her and took notes, Panzee undoubtedly wanted to do the same. At the Lola ya Bonobo sanctuary in the Democratic Republic of Congo, the bonobo Mimi played at dictation, tracing signs on a sheet of paper when asked to write. The chimpanzee Moe, which St. James and LaDonna Davis thought of as their own son, appeared to have learned how to write his own name.

The great apes thus appropriate the socialities, the savoir-faire, and the objects of those that surround them, actualizing latent abilities according to the opportunities they are afforded. Some use iPads; others have learned how to swim and even dive. Raised by Jill and Brad James in Malden, Missouri, the chimpanzee Cooper was introduced to swimming pools from his infancy. He jumps in the water, swims, and dives. He is also able to breathe using a regulator and air tanks. He never appears to be afraid, which is remarkable given that it was long thought that all great apes feared the water. Two scientists, Renato Bender (University of the Witwatersrand, Johannesburg) and Nicole Bender (University of Bern, Germany) have studied the case of Cooper. They have published about his extraordinary ability to voluntarily control and suspend his breathing while diving, although it is surprisingly little known in the media. They also observed the cross-fostered male orangutan Surya, which was living in a private zoo, the Myrtle Beach Safari in South Carolina. While Cooper uses a technique that resembles the breaststroke, Surya alternates his left and right hands when he swims, his movements like those of a frog. The Benders have found other well-documented cases of ape swimmers.[19]

Human Partners, for Better or for Worse

At the heart of these communities of humans and apes there arise new forms of proximity as well as very strong ties revealed by certain symbolic acts that signal the status of the great apes as belonging to the larger human family. Jane Goodall announced the death of David Greybeard, a chimpanzee, in the obituary column of a British newspaper. Similarly, when Austin, a chimpanzee at the Language Research Center, died in 1996 at the age of twenty-two, researchers thought it inappropriate to use his cadaver for biomedical research, as was the custom at the center. He was cremated, and his ashes were scattered in the woods where he liked to walk.

The integration of great apes into human environments also extends to their sexual preferences. It happens, for example, that certain orangutans raised by humans are interested in women. At Camp Leakey, a rescued subadult male named Gundul calmly and deliberately raped the cook in 1975. Biruté Galdikas, founder of Camp Leakey and an expert on orangutans, explained this behavior by the fact that the orangutan had always lived among humans. Great apes raised in these conditions are, in effect, frequently sexually attracted more to humans than to other apes, as was the case with Lucy the chimpanzee. As were many employees of the camp, the American actress Julia Roberts was threatened by the male orangutan Kusasi during her visit to Camp Leakey in 1996. These sexual assaults seem to confirm the appropriation of a properly human identity, the image of their adoptive mother being imprinted on these primates as an indelible reference. Showing proof of a remarkable syncretic ability, certain male apes prefer women wearing fur coats. Others are never attracted by females of their own species.

Apes That Consider Themselves Human

Apes that live in close proximity with us have already crossed a line; the boundary artificially erected between humans and animals has been blurred. Most of them consider themselves human and designate themselves as such. The young female chimpanzee Washoe encountered other chimpanzees for the first time when she was five years old. Her instructor, Fouts, asked her who they were. She responded in sign language, "Bugs . . . Black bugs." She obviously did not put herself in this contemptible category. Fouts said that she divided the world into "them" (dogs, cats, black bugs) and "us" (people). Having known only the world of humans, she undoubtedly lived her life as a "human." Fouts thought that when Washoe looked at herself in a mirror, she saw a human being. He went so far as to interpret her attitude toward her "species of origin" as racist.

When Keith and Catherine Hayes asked Viki (accustomed to manipulating images during discrimination tests) to classify different photographs into two piles, either of humans or of animals, she put the image of her biological father, the chimpanzee Bokar (whom she had never met), into the animal category. On the other hand, she put her own portrait (which she recognized as her own) among the photos of Franklin D. Roosevelt, Joe DiMaggio, and Dwight D. Eisenhower. In another case, Lucy was looking at images when she came across a representation that shocked her. Very confused, she asked in sign language, "What is it?" It was a chimpanzee. Fouts explained this reaction by the fact that Lucy did not know she was an ape but thought of herself as human. The female gorilla Koko, after having put on "makeup" with a pencil, looked at herself in the mirror, then touched her lips with her index finger and made the sign she often used to signify the word *woman.* In these different cases, the apes themselves shifted the boundaries. The great apes' interiorization of bits of human ethos can thus reach profound depths.

This collusion between great apes and humans is thus of a different order than that between humans and their pets because of the strong physical resemblance, similar dispositions, and a remarkable phylogenetic proximity, all of which result in an essential difference from other species. The instructors thus came to understand profoundly the ethos of the apes while accepting their irreducible mystery, which grew ineluctably with their advancing age. The interspecies discussions in human symbolic language added a crucial dimension to the relations between humans and apes. They reinforced the close ties already established between them. These ties had, moreover, constituted a decisive factor concerning the learning not only of language (which is highly social) but also of an entirely new way of life for the primates. These were ties totally absent in the (non)relations between scientists and users of Skinner's boxes, as in the lives of laboratory rats.

The Becoming-Animal of Humans, the Becoming-Human of Apes

The particular orientation of the "way of being in the world" of primates living in human family structures is notably marked by what I would call a "becoming-human" of great apes. This notion echoes that of "becoming-animal" in humans proposed by the philosopher Gilles Deleuze and the psychoanalyst and philosopher Félix Guattari in many of their works, such as *A Thousand Plateaus* or *Kafka: Toward a Minor Literature.* This *becoming-animal* of humans can take on many forms. It can be, for example, a shaman who becomes an animal by taking on its outer form, its dances, and a trance that takes him toward the forces of nature and animal spirits with which he has the power to negotiate. Those who manifest this becoming-animal do not imitate the animal, do not resemble the animal, do not identify with the animal, do not regress to an animal state: they experiment with the vast sentiment of belonging to a pack, to a group of beasts, and become a member of a particular species. This occurs, in a way, by "contagion": the ethos of this species emerges in its being as always already animal.

In Gambia, the American psychologist Janis Carter lived for seven years surrounded by ten chimpanzees. Isolated, cut off from almost all human contact, she interacted in an appropriate manner with the primates, sleeping and eating with them. As she herself said, she had *become one of them.* It was more and more difficult for her to become human again. She experienced a becoming-chimpanzee, much as Dian Fossey, at certain moments, had experienced a becoming-gorilla, or Jane Goodall a becoming-chimpanzee. Humans themselves are also capable of integrating the way of being in the world of another species. The female chimpanzee Lucy went in the opposite direction: she demonstrated a becoming-human surprising in its breadth and depth. She did not like the presence of other chimpanzees: she avoided them and called them "dirty animals" in sign language. Despite all the efforts by Carter to wean her from using ASL, Lucy continued to communicate in sign language with Carter as well as with another talking ape, Marianne. She remained faithful to human behavior and language until the end of her life.

All in the Family

How do these traces of human ethos inscribe themselves in the core personality of great apes? An essential characteristic among all species of the family Hominidae (orangutans, gorillas, chimpanzees, bonobos, and humans), as well as their ancestors, is an extraordinary aptitude to organize themselves in social groups. This aptitude was crucial for the survival of species endangered by the fact that they were less well equipped than others in size, strength, teeth, or claws. They were thus protected by creating very strong affective ties with those around them. Great apes placed into families were exclusively surrounded by humans, who became their substitute parents, their playmates, their friends. The apes thus took humans as their role models, as they would have with their mother or members of their troop: the desire to be and to act as humans, veritable "charismatic leaders" for them, grew deeper with the months or years of cohabitation.[20]

What is more, attentive to the expectations of their companions, they manifested an extreme sense of goodwill in their readiness to make a common world with them, showing a natural inclination to slip into our universe.[21] They thus integrated a certain number of traits, appropriating behaviors that interested them as well as those they needed to live in a human world from among the opportunities offered by the circumstances in which they found themselves. It was the primates themselves that wanted to eat at the table, dress themselves, read magazines, make tea, tie knots, or paint; driven by a strong social emulation, the apes imitated those around them.

The learning of these ways of doing begins by the intense observation ("peering") of each gesture to be copied. The mirror neurons play a decisive role in this process. When a human partner executes an action, certain of his or her very specific neural zones are activated in conjunction with this precise activity. Those same neurons are activated "in mirror" in the observing primate, even though he or she does not engage in any action. The neurons that would be mobilized in an attempt to repeat the action are thus already engaged at this step, which constitutes a certain form of "mental experience" of the series of movements to be assimilated. This mechanism is not only a key element in learning by imitation for a number of species, among them primates, it is also a fundamental element of social cognition. Relying on this intense observation, the apes then attempt to reproduce the same gestures. After a number of tries, they are able to string together the correct movements. Intensifying their practice along with the progression of their abilities, they finish by appropriating different techniques, eventually becoming experts. The behaviors and savoir-faire are thus acquired progressively, then made *Stiftung,* a firm foundation, constituting a core element that directs the apes' way of being in the world.

These apes are not, however, split between two "cultures," oscillating between one and the other. This concept, revealing a dualist way of thinking and requiring a radical separation between a *human culture* and a *primate nature,* is inadequate. The fact of being *ape* or *human* does not go back to a monolithic state that can be encompassed in a definitive characterization. Regarding apes that live in close contact with humans, I use the concept of "consilience." Its primary meaning is that of association, of union, which could be applied to communities of humans and apes in which they share their worlds, their habits, and their objects, creating a "common world" all the more rich in that it is done between hominids.

This process is made possible by the extraordinary plasticity of humans and apes, particularly during early childhood, at both the physical and cognitive levels. It is supported by certain dispositions proper to these species, dispositions marked by openings and limitations, oriented by physical and psychic structures. All individuals in this way construct a world according to their own possibilities. Joining in a process of singularization, they make their own diverse social modalities, behaviors, knowledge, ways of doing, ways of living, and techniques, inventing themselves from these available "ingredients" and from occasions arising from their surroundings. The movement is double. Humans as well are overcome by different degrees and forms of "consilience" in the natural world and in communities that combine humans and great apes. Donald, the Kelloggs' son, was witness to this.

A Common World, Aesthetic Sense, and *Funktionslust*

These communities have also brought to light a common sensibility across the subtle experiences shared between humans and primates. When the rescued chimpanzee William looked at the moon, his keeper, Stella Brewer, sat next to him. He held her and they continued to look at the moon for a long while. Lyn Miles said that when Chantek accompanied her in the hills near the University of Tennessee above Chattanooga, they admired the sunset together. The young orangutan contemplated the stars. On one of these outings, the moon seemed enormous and impressive. Chantek, in the arms of the person he calls "Mother-Lyn," looked at it intently. He turned to Miles and asked in sign language what it was. Miles attributed this question to the fact that Chantek was really astonished by this fascinating image of a part of the universe and wondered about it. In their natural environment, primates also look at sunsets.

These different examples would seem to give witness to a common openness to the world and a real curiosity about it. This essential commonality can also be seen in what we call an "aesthetic sensibility" and in *Funktionslust,* shown by primates when they engage in certain activities. The Belgian philosopher Marc Richir takes up the phenomenological notion of Funktionslust, defining it as the specific pleasure felt by certain living beings when they are doing what they know how to do. Funktionslust (from *Funktion*—function, activity—and *Lust*—joy, pleasure, desire, appetite, covetousness) is thus the joy of doing what one knows how to do well. Great apes that paint display a genuine aesthetic sense and a Funktionslust, a serious and jubilatory pleasure tied to the joy of practicing a skill that interests them.

It is the same when the orangutans Dee Dee (Lowry Park Zoo, Tampa) Chantek (Atlanta Zoo), and Wattana (formerly of the Ménagerie at the Jardin des Plantes, Paris) create assemblages, weavings, or complex knots without ever having been encouraged or rewarded. Certain apes are able to tie knots, which had long been thought of as being something proper only to humans. When they engage in this activity, the primates display great creativity. The laces and ribbons given to Wattana thus form expressive materials, the choreography of gestures necessary to tie the knots being themselves part of this expression. This working with different materials, a visible encounter between the animal and the opportunities offered in its surroundings, gives rise to creations of unexpected complexity. These creations also show the personality of the primates, each one having his or her own style.

In the process of learning, Funktionslust, moreover, plays the role of accelerator: the pleasure of doing encourages the practice, this practice in turn reinforcing competency, leading to even more intense pleasure. Little by little, expertise develops. Neglected by the scientific world (mainly because of the difficulty of describing and testing these ideas), the aesthetic sensibility, the inventiveness, and the Funktionslust observed in certain species are without a doubt powerful engines of "cultural" evolution.

The Adaptability of Captive Primates

Great apes that live in zoos and sanctuaries also display certain forms of becoming-human. In these places, we can often observe behaviors that astound us because we do not expect them from animals "poor-in-world," "living without," the pure and simple state of life proposed by Heidegger. William Hornaday, the director of the New York Zoological Park (now the Bronx Zoo) at the beginning of the twentieth century, described the expertise of many of the captive primates in a work dedicated to the intelligence and behavior of wild animals.[22] He wrote of the orangutan Rajah who, taught by the keepers, learned without any difficulty to eat using utensils,

Rajah the orangutan learned to ride a tricycle in three lessons.
(From William T. Hornaday, *The Minds and Manners of Wild Animals,* 1922)

serve himself tea, dress and undress himself, use keys, hammer nails, and ride bicycles and tricycles. He was also able to light and smoke a cigarette.

Another orangutan, Dohong, developed faculties of invention and a surprising mechanical genius. Notably, he discovered the principle of leverage and applied it to different situations. In the heart of Paris, at the Ménagerie of the Jardin des Plantes, the female orangutan Nénette planned for several days how to unscrew the bolts of her cage. Her son Tubo cleaned the windows of his cage with a damp rag. At the Atlanta Zoo, the gorilla Billie B. adored certain television programs. He especially liked Luciano Pavarotti, the Italian tenor. Although he loved to be outside, he refused to go out if his favorite singer was on TV. At the Antwerp Zoo, the gorilla Victoria drew crisscrossing curves and lines on the window of her enclosure with bird excrement.

Captive primates are thus able to find for themselves in their surroundings the necessary materials to develop certain personal activities: tying knots, drawing, handiwork, making things. Within zoos great apes are also led to adopt certain of our social codes, as, for example, the female bonobo Hermien when she was living in the Planckendael Zoo in Belgium. Hermien developed the habit of "smiling" to greet the researchers when they appeared in the observation window of the laboratory. Showing teeth, however, is normally a sign of aggression among primates. Hermien nevertheless succeeded in transmitting this form of civility to her group. Isabelle, the daughter of Victoria, made as if to speak, moving her lips as she saw her visitors do, when she was in front of an audience.

Barter and Gastronomy Among Apes

In another register, the orangutans in the Ménagerie of the Jardin des Plantes perfected a particular tactic, based on bartering, to obtain sweets. Always in search of an interesting occupation, they regularly unscrewed the nuts and bolts of their cage, which worried the keepers, who feared these objects might be used inappropriately. The keepers got into the habit of getting the bolts back in exchange for biscuits. The primates, who never needed much coaxing, perfectly understood this process. Moreover, they made sure to give the bolts back one by one, keeping the others carefully hidden, in order to get even more treats. They thus learned the barter system.

At the zoo and aquarium of the Casa de Campo in Madrid, the chimpanzees learned to "prepare" meals. Linda, a female chimpanzee who arrived in 1992, had had her teeth taken out by a former owner. She developed the habit of scraping her food against the rough wall of the interior of her enclosure, then licking the pulp and juice obtained by this method. The process took from ten seconds to fifteen minutes for each fruit. The other chimpanzees began little by little to imitate this technique of making purée or juice from the fruit. In 1996, 450 hours of observation showed that apples, carrots, oranges, mandarins, tomatoes, cooked potatoes, and lemons were prepared using this method by different members of the group. Linda thus not only perfected a culinary technique, she also transmitted this new knowledge, which thus became communal.

Apes That Walk Upright

In appropriating the ways of doing and of living that are properly human, the great apes show a certain degree of liberty in relation to modes of existence that biologists and philosophers long believed to be fixed, dictated by instinct or the laws of nature. A final example shows the extent to which adaptability is at the heart of their way of life. Nonhuman primates are all quadrupeds. Some monkeys and apes, however, move about on two feet, their behavioral elasticity being supported by a degree of flexibility in their bone structure. Their skeleton, in principle adapted to walking on four limbs, seems in effect to reshape itself as a function of biomechanical changes induced by walking upright from an early age. Raised by humans and thus surrounded by a majority of bipeds, they have adapted their means of locomotion.

This is true for the female gorilla Paki and the orangutan Teak, both residents at the Louisville Zoo. These primates show the importance of sociocultural emulation in an area that had been thought of as strictly biological: locomotion. Within the communities of humans and apes, the young primates are in effect pushed to stand upright and walk. As parents do for their children from a young age, their human partners encourage them by holding their hand and congratulating them upon their first steps. The upright position allows, moreover, richer interactions with humans, which are now possible face-to-face. A number of apes in the entertainment industry also walk upright on two legs.

One ape took this aptitude even further: Bubbles, a young chimpanzee, was the companion of Michael Jackson from 1986 to 1992. He lived with the star, ate at the same table, and accompanied him on tour, sleeping in palaces, and was even able to mimic the famous "moonwalk" dance step Jackson created. They were both immortalized in a large ceramic sculpture by Jeff Koons in 1988. This type of adoption had famous precedents: Elvis Presley also had a chimpanzee, named Scatter, as a companion, as did the French singer Léo Ferré, who took a liking to the female chimpanzee Pépée.

Thus, great apes that live in zoos or sanctuaries also slip into human worlds: they are subject to human living places, objects, habits, socialities, rhythms, medicine, and dietary regimes. Tied to curiosity, a crucial neotenic quality, the attraction primates have for novelty seems heightened in a captive environment, which is much less rich in opportunities than their native tropical forests. This interest in the new seems to constitute a mechanism for staying sane, which undoubtedly allows the primates to limit the loss of abilities and behavioral rigidity on condition that their place of captivity offers them the occasion to exercise at least part of their capabilities. Apes that live in close proximity with humans are exceptional ethologists. They observe humans as much (if not more) than humans observe them, and then they try to imitate the behaviors that interest them. Expressing essential qualities and actualizing different dispositions proper to a "common hominid base," they are thus led to experience certain forms of becoming-human linked to a fundamental part of their being: their ability to adapt.

Great Apes in Popular Culture

Chimpanzees have been used in the entertainment industry since the beginning of the twentieth century. Circuses, music halls, theaters, and traveling shows exploited their aptitude as public entertainers as well as the devoted following they inspire. Sellers of exotic species opened stores. Handlers specialized more and more in their training. Movies and the press, and later television and advertising, used chimpanzees in turn. Sometimes raised to the standing of actual stars, chimpanzees have constituted an important part of North American popular culture. For generations of Americans they have become familiar, terrifying, or endearing heroes in the form of different characters indelibly inscribed in people's minds, for example, Cheeta or Curious George. A number of other species of great apes figure in Hollywood productions (for example, the orangutan Clyde, Clint Eastwood's partner in *Every Which Way but Loose*) and popular literature. These ape stars are in some ways allowed to develop different forms of familiarity and proximity with spectators. They also give rise in the public mind to the erroneous idea, however, that chimpanzees and orangutans are no longer among the species in danger of extinction.

Ham the Astrochimp

In the spring of 1959, during the Cold War, NASA selected one hundred test pilots to participate in the "space race" between the United States and the USSR. The same year, sixty chimpanzees were enrolled by the U.S. Air Force and placed in the primate facility of the Holloman Aerospace Medical Center in Alamogordo, New Mexico, in order to test manned space flight; scientists used them to test the deleterious effects of these flights on humans. The resistance of rhesus macaques to g-force had already been tested over a number of missions at the end of the 1940s.

Among the chimpanzees, Ham, a male captured in Cameroon and raised in Florida, would have an exceptional destiny. While Alan Shepard Jr. and Gus Grissom were the first Americans sent into space on a suborbital flight on May 5, 1961, they were not the pioneers of this kind of exploit. Only three and a half years old, Ham succeeded in doing it several months before them. He was among a group of six chimpanzees selected and brought to Cape Canaveral. Trained by Tony Gentry, these apes showed themselves to be highly capable, their results surpassing those of untrained humans. Gentry had an exceptional reputation, having notably trained the chimpanzee Jiggs, who played the part of Cheeta.

During his training, Ham had to wear a straitjacket and remain for hours in a special seat that restrained his legs and upper arms. He learned a series of maneuvers by operant conditioning: he was given electric shocks to the soles of his feet if he made the wrong response and rewarded with banana candy if he acted according to instructions and within the designated time. On January 31, 1961, the young chimpanzee was put into a Mercury Redstone rocket at Cape Canaveral. Ham, forty-seven inches tall and weighing thirty-five pounds, fulfilled his mission perfectly. When he was recovered in the Atlantic after a sixteen-minute flight that reached 150 miles above the Earth, he was terrorized. The public, however, interpreted the expression of fear on his small face as a large smile. He thus became not only the first chimpanzee but also the first hominid (a group that also includes humans) to be sent into space and come back alive; Yuri Gagarin made his flight two months later.[23] Celebrated as "Ham the Astrochimp," he was immortalized on the cover of *Life* on February 10, 1961. He was then transferred to the Washington Zoo, where he lived, isolated, from 1963 to 1980. He ended his days at the Charlotte Zoological Park in Asheboro, North Carolina, on January 17, 1983, at twenty-six years old, which is well below the normal age for chimpanzees in captivity.

Enos, Hero of Space

After Ham, it was the turn of another chimpanzee, Enos, to prepare for the mission of John Glenn, who would be the first astronaut to fly in orbit around the Earth, on February 30, 1962. Three months before Glenn, Enos rode aboard the Mercury Atlas rocket. On this day, November 29, 1961, the chimpanzee, about five years old, behaved like a true hero: despite the malfunctions of the electronic system, he conscientiously performed all the tasks he had learned during the entire flight of over three hours. Because of the malfunctions, moreover, he received electric shocks even when he was doing the correct maneuvers. While humans had demonstrated great intelligence and extraordinary techno-scientific abilities to be able to send rockets into space and later to walk on the moon, they had evidently largely underestimated the cognitive capacities of their close relatives as well as their indisputable willingness to collaborate with NASA technicians. Showing himself to be more intelligent than the machine, Enos demonstrated that he was careful to successfully complete his mission and that he perfectly understood what was expected of him. He was in no way, however, thanked for his heroism; he died of severe dysentery, alone in his cage, eleven months after the flight.

To make matters worse, and despite the knowledge that was beginning to build up about them (social competence, cognitive capabilities, emotional life, technical aptitude), the great apes at Holloman were later considered "surplus materials" for which a new use had to be found. In 1965 the air force decided to "recycle" them and their descendants, sending them to biomedical laboratories. It was an ignominious way to treat these "chimponauts" that had greatly contributed to the success of a major project: nothing less than the victory of the United States in the race to the moon, announced in 1961 by President John F. Kennedy.

Apes Misbehaving

The spaces where great apes and humans live together would seem to be idyllic. They actualize a form of fundamental harmony, mutual understanding, and shared language. Who has not dreamed of adopting a baby chimpanzee? However, this beautiful harmony is imperiled once the adopted apes reach puberty. They then become very strong and potentially dangerous, and they seek out sexual partners. This is why primates included in experimental psychology programs or language research are extremely young. They present no danger to the scientists and are malleable and controllable. As for the apes in the entertainment industry, the younger they are, the more the public finds them endearing. Audiences are deeply moved by their resemblance to humans as well as their childlike traits. Businesses that use them see increases in attendance, sell related merchandise, and expand their audience, procuring substantial profits for the movie industry, circuses, and trainers.

Seen as "unusable" once they enter adolescence, because of their dangerousness, the apes generally are put in a cage out back, waiting to be sold. Those who have been placed with families have to be sent away, thus brutally ending the life they shared with their human companions. In addition, apes raised by humans do not have the appropriate social skills for their species. Having taken a different path in their development, in a world not their own, they have never had the opportunity to acquire the behaviors, knowledge, social rules, and protocols of interaction essential to living in harmony with other apes. Taken from their mothers at a young age and raised by humans, they have adopted incongruous habits in the eyes of other primates: they do not know how to be apes. Their integration into a group therefore poses enormous problems.

For a long time, these primates had no hope of a peaceful retirement. Today there are over ten sanctuaries in the United States that take them in. Taking care of an ape, however, is extremely expensive. The estimate for the lifetime care of a single ape is around $1 million, without any possibility of recovering the cost. The sanctuaries have a self-imposed rule not to exploit their residents, and they are not, in principle, open to the public. The search for funding is therefore a daily struggle. The dramatic situation of all these primates that have been placed in the service of humans, forced to enter into our world, requires that we recognize the depth of our responsibility toward them.

Great Apes and the American Way of Life

The becoming-human of great apes is one of the most fascinating aspects of the resemblance between humans and primates, the marker par excellence of the depth of our continuity. Distinguished by an unparalleled originality, the "apparatus" in which this becoming-human is expressed has allowed us to understand more fully "what it means to be an ape." Going well beyond the famous phrase "Monkey see, monkey do," the primates have shown exceptional behavioral flexibility as well as an ability to make human worlds their own. They have gone so far as to appropriate aspects of human ethos and to think of themselves as human. This becoming-human of primates is, moreover, profoundly marked by the spirit of particular times and societies: clown and public entertainer during antiquity and the Middle Ages, tea drinker in a frock coat during the eighteenth century, cigar smoker putting coal in the stove in the nineteenth century, object of research in the twentieth.

After World War II, the entry of apes into human families as full-fledged members allowed for the expansion of the range of possibilities they had been given. They had the opportunity to discover novel activities, adapt to new habits, play with objects that were fascinating for them, and to fit (to a point we did not expect) into ways of life that deeply interested them. Was not the chimpanzee Lucy a perfect representative of North American mass consumer culture in the 1960s? She read popular newspapers, drank soda, watched television, and sipped cocktails among friends. She was so familiar with the industrial technology available to the 1960s housewife that she used the vacuum cleaner as a dispenser of orgasms on demand. Daily elements of the American way of life, spread throughout postwar Europe, held no secret for her. She missed them terribly, however, on her return to Africa, a universe completely foreign to this "American" cover girl for *Life* magazine.

But the essential community shared by apes and humans was not limited to this becoming-consumer. It was also filled with aesthetic sensibility, inventiveness, Funktionslust (elements often neglected by the scientific world, although they are determinant for learning and cultural dynamics), and a common openness to the world, notably revealed by the emotions shared looking at sunsets.

MY FIRST GORILLA.

The Franco-American explorer Paul Belloni Du Chaillu killed his first gorilla in Gabon.
(From Belloni Du Chaillu, *Explorations and Adventures in Equatorial Africa,* 1861)

SIX

Conquering the Field

Pioneers, the Quest for Origins, and Primates

BETWEEN NOVEMBER 1884 AND FEBRUARY 1885, the European nations divided up Africa at the Berlin Conference. New needs prompted by the industrial revolution stimulated colonial expansion. Infrastructure, trading posts, religious missions, outposts, and means of transport all were organized to facilitate the exploitation of the lands being colonized, opening up new opportunities for explorers, hunters, and businessmen. Scientific expeditions were sponsored by several large American and European institutions.

At the time, scholars still had very limited knowledge about the lives of great apes in their natural environment. They knew little about their diet, nothing about their habits, and could in no way imagine the sophistication of their social life. From the beginning of the twentieth century until just after World War II, scientists spent their careers within laboratories, museums, and universities, supposed to be the only adequate places to do "good science." Except for several rare individuals at the head of large expeditions, biologists did not observe wildlife in locations deemed dangerous. Travelers still died of malaria, amoebic dysentery, sleeping sickness, and blackwater fever. They were attacked by large cats or elephants—and sometimes by native people. Guides and porters on occasion abandoned expeditions.

What is more, studies of animal behavior were considered anecdotal and not very serious: "In zoology, anatomical descriptions have long constituted the principal distinction between scientific literature, where they abounded, and popular literature, where they were more restrained. Also, the descriptions of 'animal customs' seemed to belong ipso facto to popular literature."[1] The ways of life and the societies of primates were therefore observed in laboratories and zoos, where the groups were formed artificially. These were, however, the only studies said to be rigorous, since they were conducted under controlled conditions.

Before the twentieth century, only a few naturalists went to observe primates in their natural environment. In the 1920s and 1930s, three of Yerkes's students would attempt scientific work in the field. Only one of them really succeeded. Japanese researchers began systematic work on Japanese macaques in 1948. They showed that the macaques displayed "cultural" behavior as early as 1954.

The First Field Expeditions

The observation of the way of life of animals in their natural habitat constitutes, paradoxically, a minor tradition in the life sciences. Before World War II, the study of primates was placed with the academic culture of museums of natural history as well as an experimental science practiced in laboratories where, since the beginning of the twentieth century, dozens of primates were used in studies of comparative psychology, as well as in anatomical, physiological, and biomedical research. When scholars participated in scientific expeditions, their mission was essentially to collect exotic species to enrich the collections of museums or universities. The observation of animals was secondary, occasional, and opportunistic. Thus, during the first systematic scientific research in northeastern Congo for the American Museum of Natural History Congo Expedition (from 1909 to 1915), Herbert Lang and James Paul Chapin would bring back more than one hundred thousand specimens, including fifty-eight hundred mammals. The collection, preservation, preparation, and shipping of the specimens left little time for other tasks. However, the two men certainly encountered great apes: a chimpanzee was among the many magnificent portraits of primates taken by Lang, an excellent photographer as well as a scientist.

Nonetheless, many naturalists (including Charles Georges Leroy, both a pioneer and a visionary) had advocated research in natural surroundings since the eighteenth century. At the turn of the twentieth century, Charles Otis Whitman argued for the necessity of long-term studies of animals' way of life as well as for an evolutionist approach to behavior. He prefigured modern ethology. Within the framework of German natural philosophy, Jakob von Uexküll created an institute for the study of place and environment in Hamburg in 1925. He advocated the concept of Umwelt ("world" in the sense of the "representative universe" of each animal) and insisted upon the importance of studying animals in their own habitats. Following von Uexküll, European ethologists promoted the study of the "natural" behavior of animals in their own environments. In addition, new questions emerged about the social organization of primate communities, food chains, and ecological niches. Later, genetics and population dynamics, sociobiology, and the central place accorded to an evolutionist perspective forced scientists to take into account the interactions between individuals as well as their relations with their socioecological environment. Field research became indispensable; the laboratory no longer sufficed.

The Malay Archipelago

Before the twentieth century, observations of great apes in their natural environment were made by sea captains, travelers, missionaries, explorers, hunters, or traders. These were seen as unreliable amateurs by both scholars of their time and later commentators. Some naturalists, however, having the status of "professionals," also took part in several voyages of exploration. This was the case for Darwin as well as Alfred Russel Wallace, cofounder of the theory of evolution, who went to Indonesia in 1854 and stayed nearly fifteen months at Sarawak on the island of Borneo. He told of his first encounters with orangutans (called *mias* by the Dayaks) in *The Malay Archipelago,* a work he dedicated to Darwin. Wallace had occasion to observe their behavior, feeding habits, and nests (built each night) for more than one year and in different environments. He said that the apes were in competition with the Dayaks since they fed on the same fruit. Wallace also stated that the mias attacked humans: when they saw him, they broke off branches and threw them at him. It must be remembered that Wallace was above all a hunter and collector, and that his initial reaction upon seeing these animals was to get his gun and shoot. Observation came second.

Tied to the way in which Wallace saw relations between humans and apes, the orientation of his work gave to these Asian "men of the forest" an entirely different image than that which would later be promoted in the articles by Biruté Galdikas in *National Geographic.* Reinforcing the collective imagination, Wallace stressed the dangerous aspects of the ape. In the opening pages of *The Malay Archipelago* he illustrated his work with an engraving of an orangutan with its "fangs" planted in the thigh of a local person, perpetuating the traditional representation of the orangutan as the incarnation of savagery and violence. Wallace nevertheless also described certain customs of orangutans, such as their way of moving through the canopy, which he compared to that of gibbons.

He would have a further occasion to get to know the species. After he killed a female, he found a baby orangutan hanging on to the body. He took care of the infant, feeding it, bathing it, and making it do "exercises" to work its muscles. He even gave it a baby monkey as a companion and observed their play. He was amused by the mimicry of the young orangutan, which died after three months. Wallace said that he was sad to have lost his "little pet." He nonetheless preserved the skin and skeleton. He returned to England in 1862 after eight years away. He had collected more than 125,000 specimens, including 310 mammals. Among his take were representatives of many species of monkeys as well as orangutans of different ages, including large adult males. Some of these specimens are conserved at the British Museum.

"My Little Pet Orang"

Several years later, in 1877, the taxidermist and zoologist William Hornaday left for India and Ceylon. One year later, he arrived on the Malaysian peninsula. He spent a number of months on Borneo, collecting specimens. Expert taxidermists were at the time highly esteemed since they were capable of preparing and preserving specimens destined for museums during expeditions. Like Wallace, Hornaday went to Sarawak and studied orangutans.[2] He observed their nests and the way they were built. He noted that the orangutans first chose a fork in a tree and then filled it with branches. They completed this construction by surrounding themselves with branches and leaves. Hornaday noted that orangutans never used the same nest twice. He further remarked that they were often solitary, forming small groups with a maximum of three individuals, composed of mothers with their offspring. Only a single young was raised at a time, although juveniles remained a long time with their mothers. Studying their diet, Hornaday found that their preferred fruits were durians, mangosteens, or rambutans. He refuted the notion that orangutans walked upright.

One day, the Dayaks brought him a young male orangutan, extremely intelligent, which would stay with him for four months. Hornaday said that the infant (which he called "my little pet orang") seemed to think of him as a foster father; Hornaday loved the young ape as if it were his son. Hornaday noted that the orangutan covered itself with straw and cloth when it rained, leading him to believe this could be a habit among orangutans in their natural environment. He also took care of three other baby orangutans, aged from seven to eight months. He described their ways of being, their demands, their impatience, their anger, and their complaints, comparing them to those of infant humans. Hornaday defended the notion of individual diversity and a personalization of primates, noting as well that facial characteristics varied greatly from one ape to the next. He nonetheless fulfilled his mission as collector by killing forty-three orangutans, including a pregnant female. He carefully measured them all, including the fetus. His observations followed in the naturalist tradition, like those of Wallace. Upon his return to the United States, Hornaday became the first director of the New York Zoological Park (today the Bronx Zoo), which opened in 1899. He also was one of the first great American conservationists.

African Adventures: Paul Du Chaillu and Richard Burton

Such "professional" naturalists were, however, somewhat rare. It was an "amateur" whom scientists deemed unreliable who made some of the most important observations on great apes in the mid-nineteenth century, the Franco-American hunter and collector Paul Belloni Du Chaillu. Joining his father after studying taxidermy in France, Du Chaillu first got to know Africa in 1848 at the age of seventeen. While working with his father, a prospector and a trader, he studied with the missionaries on the western coast. After a brief time in Paris, he returned to United States in 1852. In 1855, he left again for Africa, exploring both inland and along the coast of Gabon. He collected a number of specimens, some of which he sent to the Philadelphia Academy of Natural Sciences, until 1859. He also hunted gorillas for the first time. Du Chaillu would travel to Gabon again from 1863 to 1865, encountering many primates and bringing back specimens, including several gorillas and chimpanzees.

He was one the first foreigners to have seen western lowland gorillas in their natural environment. He initially recounted his experiences in a nuanced and realistic manner. At the demand of his editors, however, he rewrote his adventures, now depicting gorillas as terrifying beasts in order to satisfy readers' sensationalistic appetite. He said that gorillas recalled the hybrid beings described by ancient writers, hairy and savage, a dangerous mixture of bestiality and humanity. He lost all credit in the eyes of scientists, many of whom accused him of lying.[3] Some even suspected him of never having been to Africa.

Du Chaillu did not limit himself, however, to these spectacular descriptions that seemed taken from popular literature. He debunked a number of stereotypes and taught his readers that gorillas were timid, nomadic, and lived mainly on the ground. He also observed a number of chimpanzees and even adopted a baby of the species "Nsiego-Mbouvé." He related that the infant chimpanzee had spent a long time looking at his dead mother, sensing she had undergone a change; he touched her, sadly crying out. Named Tommy, the chimpanzee quickly became attached to humans at the camp. Du Chaillu said he was intelligent, loved the society of men, and was perfectly adapted to "civilized life." He liked being held, and he carried around with him the cushion upon which he slept. The young chimpanzee died after five months. Du Chaillu also took care of several young gorillas.

During the same period, a number of other adventurers visited the land of the great apes, which were prestigious trophies for the "Great White Hunters." In 1876, the explorer Richard F. Burton published *Two Trips to Gorilla Land and the Cataracts of the Congo* about his travels in "the land of gorillas." He confirmed some of Du Chaillu's observations and refuted others, for example, their fiery eyes and terrifying cries, and seeing the gorilla as a hybrid creature, part man and part animal. Burton learned from local peoples that the coat of adult males was "silvered." He also showed that certain beliefs at the time were false, such as that gorillas walked upright.

"Black Monkeys"

In 1902, the German explorer and army captain Oscar von Beringe was most likely the first Westerner to encounter a living mountain gorilla (Colonel Ewart S. Grogan had seen only a skeleton in 1898). Rwanda was at the time part of a German colony, German East Africa. Accompanied by Dr. Engeland and Corporal Erhardt, von Beringe was charged with visiting outposts in the volcano region (situated between the current Uganda, Rwanda, and the Democratic Republic of Congo) previously explored by the Englishmen Speke and Grant. Accompanied by a small team, von Beringe reached an altitude of 10,400 feet up Mount Sabyinyo when he spotted some "black monkeys" on the peak of a volcano. Two primates were killed, though only one was recovered. It measured five feet and weighed two hundred pounds. It therefore could not have been a chimpanzee.

Having no scientific training, von Beringe nevertheless thought it could not be a gorilla, since none of that species had been found in the region up to that time. The remains were sent to the Berlin Museum. The German zoologist Georg Friedrich Paul Matschie described it as a new subspecies of gorilla and gave it the official name of *Gorilla beringei beringei,* in honor of the German officer, in 1903.[4] Von Beringe tried to get to know more about these strange animals. He followed their tracks through the vegetation and examined their nests, excrement, and other traces. He was thus able to confirm the presence of the large primate at various sites in the Virunga Mountains. The Grauer's, or eastern lowland, gorilla was first identified by Rudolf Grauer in 1908.

Ben Burbridge, *The Gorilla Hunt*

In the 1920s the American hunter, collector, filmmaker, and photographer Ben Burbridge also went in search of gorillas. His documentary *The Gorilla Hunt* (1926) is a pioneering film that depicts the great apes in their natural environment. In *Gorilla: Tracking and Capturing the Ape-Man of Africa* (1928), Burbridge goes back and forth between empathy and stereotype, as had Paul Du Chaillu, whom he mentions in the work. There is empathy when he denounces serious misconceptions about the species and brings to the fore the personalities of gorillas of all ages, males as well as females, that he encountered. The stereotypes come through when he uses certain clichés, for example, the "man of the forest," and describes them as brutal and territorial, incarnating the Stone Age. He places the accent on the bravery of the hunter confronting the most terrifying beasts in the heart of Africa.

Burbridge was among the first professional hunters to furnish young gorillas for European and American zoos, including a mountain gorilla for the Antwerp Zoo in Belgium and the famous Miss Congo (also a mountain gorilla) studied by Yerkes in 1925.[5] To the violence of the hunt he added the brutality of the capture of the young taken from their mothers. Despite the fact that he was above all a businessman (dealing in specimens, film, and books), Burbridge greatly interested Yerkes, who wanted to profit from his experience in the field. Georgina Montgomery asserts that their collaboration represents a good example of relations being established between scientists and laymen, professionals and amateurs.[6]

Martin and Osa Johnson, *Congorilla*

Reports of great apes thus gradually became more numerous. In the 1930s, a new version of the gorilla, peaceful and attached to its family, emerged. This was most notably seen in the dioramas created by Carl Akeley in 1936 for the Hall of African Mammals at the American Museum of Natural History in New York. This representation of gorillas, however, came at the same time as another, dominant and infinitely more popular, presented in the film *King Kong* (1933). Filmmakers Merian C. Cooper and Ernest B. Schoedsack chose to make their gorilla a creature that embodied, above all, savagery. The *Beast* occupied metaphorically the place of local peoples who, in the colonial imagination, were driven by desire to possess the *Beauty*. The forced alliance between the "white woman" and the "black man" symbolized the feared mixture of incompatible natures.

Scientific data, reports from the field, and popular stories (having long taken root in people's minds), were contradictory. They cast an ambiguous image of the gorilla, of which the film *King Kong* is a reflection. Other films showed a more amiable hero; these were ignored by the general public. Made by the American explorers Martin and Osa Johnson, the film *Congorilla: Adventures with the Big Apes and Little People of Central Africa* (1932) had a much more fleeting effect on the popular imagination. Nonetheless, the Johnsons wanted to offer to viewers an authentic vision of Africa. Martin (who had traveled with Jack London) and his wife Osa were amateur naturalists fascinated by the people and fauna of Central and East Africa. With the support of the American Museum of Natural History and George Eastman (founder of the Eastman Kodak Company), they filmed segments of the Akeley-Eastman-Pomeroy Expedition in 1926, in which Carl Akeley and his wife Mary participated.

The authors of numerous works, the Johnsons filmed *Congorilla* (one of the first films with sound shot in the Congo) during their fourth voyage to Africa, from 1929 to 1931.[7] In the last part of the documentary, viewers see the giant primates in the natural environment in the hills of Alumbongo. Peaceful and debonair, the gorillas eat, play, and rest. The Johnsons' film shows an idyllic Congo where humans and animals live in Edenic harmony. The married explorers, however, did not hesitate to break this harmony: they captured two juvenile gorillas and bought another from local people, then took the three infants from their native Africa and brought them to the United States, without the authorization of Belgian colonial authorities. The Johnsons nonetheless have the merit of offering in their film a

Osa and Martin Johnson with an orangutan sitting on their amphibious plane at Camp Abai, Borneo, circa 1935.
(Photograph courtesy of the Martin and Osa Johnson Safari Museum, Chanute, Kansas)

new image of gorillas, finally freed from the stereotypes projected on the species. They later traveled to Asia, and found that orangutans were not the dangerous wild animals described by other travelers.

Conquering the Field

In concurrence with these different field experiences, research was also being conducted by "amateur" naturalists who adopted a scientific approach. Reconciling his activity as a trader in exotic animals with his desire to study primates in their natural environment, Richard Lynch Gardner made a number of trips to Gabon and the French Congo (today Republic of the Congo). Between 1892 and 1893, he made observations in the African forest using a protective cage that he called Fort Gorilla, similar to the anti-shark cages divers use. He showed in this way that from the outset of his work he considered the great apes potentially dangerous beasts he needed protection from. Square, measuring six and a half feet on each side, and built specially for the purpose in the United States, the famous steel-wire Fort Gorilla was painted green and covered with bamboo leaves. According to Garner, he spent nearly four months in the "furnished" cage with a chimpanzee named Moses, although some critics think it was only a few days.

After conducting research in a zoo on the language of primates, he tested their aptitude for communication in the field. He wanted to be the first to record the vocalizations of chimpanzees and gorillas in nature in order to analyze and compare them. His project was ambitious: nothing less than to shed light on the origins of language. He wanted to record the apes using an Edison phonograph, but he was ultimately unable to take the equipment into the field.

Throughout his career, Garner would see his work questioned by scientific authorities and the press. However, he stated that he observed gorillas, chimpanzees, and other species of primates on a number of occasions. He became interested in their diet, understood that chimpanzees were better climbers than gorillas, and knew that both species were nomadic. An excellent observer, Garner described their way of moving through trees in detail. Wanting to be closer to them, not only through experiments and observations but also through sharing daily activities, he adopted a number of chimpanzees (notably Moses) and gorillas (notably Othello). Some of them went with him to England (Aaron and Elishiba) or the United States (Susie and Dinah). Devoting the majority of his life to the study of apes, Garner was in a position to refute some assertions about them in the popular literature as well as make pertinent, even pioneering, observations and reflections about their intelligence and behavior. He furthermore defended Darwin's theories against those people with religious prejudices.

Also an "amateur" naturalist with a scientific approach, the South African naturalist, jurist, and writer Eugène Nielen Marais conducted studies of a particular troop of Chacma baboons for a number of years, beginning in 1905, in the Waterberg region of South Africa.[8] He showed the complexity of their societies, studying questions of social hierarchy and sexual behavior. Working within a Darwinist and functionalist paradigm, Marais also described certain games, forms of communication, and sex-specific roles within the group. He was interested in the cognitive capacities of baboons and remarked that they retained a memory of certain cause-and-effect ties (causal memory). He further showed that baboons were carnivores and that they hunted, which would later greatly interest Raymond Dart, who developed the "killer ape" theory. Although he was also considered an amateur, Marais received more recognition than Garner from the scientific community.

The Great Scientific Expeditions: Between Museum, Laboratory, and the Field

After a long phase of discovery by legendary explorers, such as David Livingstone and Henry Morton Stanley, European colonies became occupied under various forms (religious, commercial, political) and organized by an economy tied to the exploitation of beings and resources. Mandated by Western museums, universities, and institutions, scientists took part in a vast scientific mission of an inventory of all forms of exotic life in places heretofore unexplored. However, at this time of transformation in the field of zoology, which was gradually admitting the importance of species' natural environment in research, scientists began to pose new questions and study more seriously the behavior of the animals that interested them.

In 1896, a young taxidermist, Carl Ethan Akeley, took part in the British Somaliland Expedition organized by Chicago's Field Columbian Museum (today the Field Museum of Natural History). It was the first zoological expedition organized by an American museum on the African continent. It was also the first time Akeley went into the field. He fell in love with Africa. During the expedition, he brought along a camera, previously considered too awkward and useless by his colleagues, as well as the equipment necessary to set up a darkroom in the field: the glass plates used in the camera needed to be developed soon after their exposure. Akeley took more than three hundred photographs.

A Taxidermist, an Explorer

Accompanied by his first wife, Delia, and one of the most famous hunters of the time, R. J. Cunninghame, Akeley left for a second expedition in Africa organized by the museum in British East Africa from 1905 to 1907. Two years after his return, he was hired by the American Museum of Natural History (AMNH) in New York. He was already dreaming of an ambitious series of monumental dioramas that would give visitors the illusion of being in Africa. He gave his body and soul to the project.

Akeley went on three further scientific expeditions (1909–11, 1921–22, and 1926). On the first, he became interested in the behavior of certain animals, particularly elephants, while fulfilling his mission of collecting specimens. During the second two expeditions, he went to look for the mountain gorillas described by Captain von Beringe. He wanted to study them in their natural environment so that he could give a realistic description of them in books, films, and photographs. He was the first to film a documentary on gorillas in their natural setting (entitled *Our Great Primitive Cousins*) and came back with over three hundred feet of footage.

Strongly criticizing Du Chaillu, he denied the supposed ferocity of the species, describing them as peaceful vegetarians with strong family ties. He noted that groups were often composed of an adult male with several females, and that they built nests on the ground. Faced with these apparently debonair giants, Akeley had mixed feelings. A distinguished hunter, he sometimes had the feeling of being an assassin. He nonetheless killed the five animals necessary for one of the twenty-eight dioramas built for the Hall of African Mammals, which opened in 1936 (ten years after his death). When he observed the behavior of gorillas, it was often in the role of the "Great White Hunter," while he was tracking and hunting the primates as prey. He affirmed that gorillas were not aggressive animals, as had often been thought. When they charged humans, they did so for reasons of defense, protecting their families.

Observing Mountain Gorillas

In 1926 Akeley went on the Akeley-Eastman-Pomeroy African Hall Expedition with his second wife, Mary Jobe Akeley, an explorer who had studied at Columbia University. Accompanied by their guides, trackers, hunters, cooks, and porters, they set up camp at Kabara, determined to do an in-depth study of mountain gorillas in the area of the Mikeno and Karisimbi volcanoes. They were aided by a talented and experienced Congolese forest ranger, Sanwekwe, who would later also assist George Schaller and Dian Fossey.[9]

Akeley thus began a new field of investigation of primates. Unfortunately, the project was interrupted by his untimely demise on November 17, 1926. His wife had him buried at the foot of Mount Mikeno. The Belgian zoologist Jean-Marie Derscheid continued the expedition. He observed a number of families of gorillas and described their behavior.[10] He

Carl and Mary Akeley in their camp during the expedition Eastman-Pomeroy-Akeley, near Lukenia in Africa. Carl is at work on a sketch model for the Klipspringer group of the future Akeley Hall for African Mammals (1926–1927).
(Image #412183, Courtesy of the American Museum of Natural History Library)

also worked on the gorillas' spatial distribution and population. Derscheid was thus the first to conduct this kind of research in Albert National Park as part of a scientific expedition, between 1926 and 1927.[11] Charles Robert S. Pitman, the game warden of the Uganda Protectorate between 1925 and 1951, studied mountain gorillas within the Kayonza Forest in 1933 and 1934.[12] He described the fauna and flora of the region, closely studied the gorillas' nests, and explained that they built them in trees. He also made several observations of chimpanzees. Mountain gorillas would be the subjects of another investigation in 1937.

Carl Akeley and the Hall of African Mammals

Of all the dioramas made for the American Museum of Natural History, that of the gorillas of the Virunga Mountains was undoubtedly closest to the heart of the man considered the father of modern taxidermy. Through these works of art characterized by an obsession with realism, Akeley tried to capture the glory of Africa for visitors of the museum. He made use of all possible means in this work of reconstruction: specimens, death masks, taxidermy, paintings and drawings from nature, sculpture, film, photography, notes, and detailed plans.

His obsession led him, paradoxically, to set forth an idealized vision of nature, where sick animals or those without noble qualities (cowardice, ugliness, abnormalities) had no place.[13] This spirit of perfection keeps viewers at a distance while allowing them to look at this fixed representation of nature for as long, and in as much detail, as they want. The experience is one of contemplation. By presenting the great apes in family groups within a carefully reconstituted environment, he rendered them both strange and familiar to the public. He made the gorillas accessible, taking away their mystery while at the same time reestablishing the separation between humans and animals by enclosing them, frozen, behind glass.

Between the nineteenth and twentieth centuries a powerful international movement for the protection of nature developed, led by numerous Western naturalists. Worried about the future of gorillas, Akeley wanted to put in place a vast conservation program, particularly one that would create a sanctuary for the gorillas he saw as threatened. It was under his inducement that Albert National Park was created by King Albert I in the former Belgian Congo, based on the model of Yellowstone National Park in the United States. It was the first sanctuary dedicated to the conservation of nature in Africa. Today, under the guidance of Emmanuel de Mérode, the park director since 2008, Virunga National Park constitutes one of the greatest areas of biodiversity (in birds, mammals, and reptiles) in Africa.

From Laboratory to the Field

A major figure of modern primatology, the mammalogist Harold Coolidge Jr. (Harvard University) participated in the last three great expeditions conducted in the nineteenth-century tradition: the Harvard Medical School African Expedition (in Liberia and the former Belgian Congo, 1926–27), the Kelley-Roosevelt Expedition (in Asia, 1928–29), and the Asian Primatology Expedition (in Asia, 1937). The collections obtained from these expeditions contributed greatly to the growth of knowledge and field research techniques in the first part of the twentieth century. The Harvard African Expedition was the first systematic scientific expedition organized in Africa by the university. A specialist in primates, Coolidge worked on the distribution and classification of lowland gorillas in central Africa and brought back a specimen for the Museum of Comparative Zoology at Harvard. He later observed gibbons in Asia, initiating the first behavioral study of the species in its natural environment.

Navigating between ethology and anatomy, his research constantly brought out the tension between fieldwork and studies done in natural history collections, between observation of living animals and examination of dead ones, in the process learning much from the confrontation between these two worlds. Coolidge was thus able to place particular anatomical and morphological characteristics in relation to their function. He would be the first to organize classes dedicated specifically to primates, at Harvard in 1931. A pioneer of the international conservation movement (he was one of the founders of the International Union for Conservation of Nature as well as the World Wildlife Fund), he questioned the habit of killing large numbers of animals for collections. He thus embodied a historic transition that began with an epistemology proper to museums and then oriented itself toward a paradigm that included the question of conservation within the discipline of zoology. It was Coolidge who sent George Schaller, until then a specialist in birds, to observe mountain gorillas at Kabara for twenty months in 1959, eight years before Dian Fossey began her work in the same area.

Henry Raven, in Quest of Gorillas

Among the great scientist-adventurers of the early twentieth century, Henry Cushier Raven united in a single person all the various abilities of a field researcher. An expert in primates and the author of numerous articles on other species, he was, however, self-taught, and did not attend university except in the course of his career, becoming himself a lecturer in zoology at a number of institutions, including Columbia. Having benefited from training in taxidermy while gaining a naturalist's knowledge and experience as a hunter during his adolescence, he was capable not only of directing expeditions and learning local languages but also of tracking and killing animals as well as preparing and conserving them before shipping them off to the West. He was also a photographer and a competent supervisor of both the finances and the personnel of expeditions.

Exploring a number of regions throughout the world, Raven worked for many well-known institutions, most notably the American Museum of Natural History. In May 1929 he left for Congo as part of the American Museum of Natural History–Columbia University Gorilla Expedition in the Kivu volcano country. With the aid of Batwa Pygmies, known as "the keepers of the forest," he followed groups of gorillas and tried to study their behavior, although, like Akeley, he was there more in the role of a collector.[14] The scientists of the American Museum had received permission from the Belgian authorities to collect two specimens. It was from these two specimens (one of which weighed five hundred pounds) that Raven would conduct the first complete anatomical study of a gorilla. He recounted his adventures in a book, *In Quest of Gorillas* (1937), which he wrote with the curator of anatomy at AMNH, William K. Gregory, his supervisor, colleague, and friend.

Raven spent a lot of time in the field, which gave him many opportunities to make observations. He came to know the habits, diet, nests, feeding practices, and defensive behaviors of gorillas. He established comparisons between gibbons, orangutans, chimpanzees, and gorillas, and acquired an intimate knowledge of chimpanzees. During the same expedition, now in French Cameroon, he lived for a year, practically alone, with a young female chimpanzee acquired in February 1930 and named Meshie Mungkut. A year later, he brought her with him to the United States to his family home on Long Island, New York, where she lived for four years. Furthermore, Raven made meticulous and detailed contributions in mammalogy and comparative anatomy in numerous articles. A major work, illustrated with impressive engravings done under Raven's supervision, *The Anatomy of the Gorilla* (1950) was published posthumously.

Early Fieldwork

Woven through the heart of the Western collective imagination in the form of stories, films, photographs, and reports, the big scientific expeditions generated new vocations while breathing new life into the great Linnaean project of naming and classifying the earth's diversity. Popular fiction continued to have a big impact. Certain best-selling authors, however, never went into the field, such as Edgar Rice Burroughs, the creator of Tarzan, who published twenty-five books about the ape-man between 1912 and 1965. Readers nonetheless imagined chimpanzees and gorillas as Cheeta or King Kong rather than according to the descriptions given by scientists. It was time for serious research to be put conducted in order to show the public what apes really looked like.

The great primate specialist of the time, Robert M. Yerkes, went to Africa in 1929. He wanted to create breeding stations and sanctuaries there, based on the model of Albert National Park. Deeply committed to verifying and validating the knowledge about great apes gained in laboratories, he also wanted to organize long-term field studies. Determined to establish an international research program, he first visited the Institut Pasteur at Kindia in French Guinea. At the end of the same year, he sent one of his students, Henry Wieghorst Nissen, into the forests around Pastoria (the Fouta Djallon region).

Henry Nissen, an American Scientist in Africa

Having been provided with information by Rosalía Abreu, Nissen spent two and a half months in the field. He had three objectives: evaluate the feasibility of studies of wild chimpanzees, develop methodologies and specific techniques applicable to field research, and gather all the practical knowledge necessary for taking care of the primates that would be kept at the station at Orange Park. He was also (and perhaps chiefly) charged with procuring chimpanzees for the two research centers founded by Yerkes at New Haven and Orange Park under the auspices of Yale University. Inexperienced, he left for the jungle with his only methods those traditionally used by hunters and naturalists: observation blinds, traps, and tracking animals by their various traces. The blinds, however, being fixed, were of little use in studying nomadic animals. The chimpanzees ignored the food left as bait. Finally, when tracked by humans, the apes noticed them immediately and fled.

After sixty-four days in the field (forty-nine days of observation), however, Nissen was finally able to approach the primates, first to a distance of 150 feet, then 50, and sometimes even to within 20 feet. He placed himself by their nests at daybreak and observed their morning activities. He reported that the chimpanzees were highly social and lived in small troops. The time he spent with them, however, was too short to identify all the individuals observed or to understand their social organization. Nissen nonetheless stated that infants stayed with their mothers for several years and slept in the same nest as them. He also described in detail certain means of communication among the chimpanzees as well as their diet, noting that this included some small game. Reporting the observations of local populations, he also said that females distanced themselves from their troop while giving birth.[15] Fulfilling one of his principal tasks, he returned to the United States with sixteen young chimpanzees, a gift from the Institut Pasteur, for Yerkes's laboratories.

Harold Bingham and the Mountain Gorillas

Harold Clyde Bingham, another of Yerkes's postdoctoral students, was also charged with conducting a preliminary study, this time on mountain gorillas. Yerkes wanted him to verify the results gained from the study done of Miss Congo. Lacking any experience, Bingham read the reports written by the Akeleys. He also visited Rosalía Abreu's colony in Cuba. Finally, directing the joint Yale University and Carnegie Institution Expedition, he went to Albert National Park and stayed there from 1929 to 1930 with his wife Lucille. Wanting to conduct a "psychobiological" study, he first of all examined the possibilities of conducting research at the site and tried to count the gorillas. He was often accompanied by a number of porters, in the tradition of former expeditions, which frightened the primates and kept them at a distance.

Bingham nevertheless managed, little by little, to follow certain groups. He even succeeded in staying with one of the groups for a period of one hundred hours. He examined the nests (both for night and day use) made by the apes, the composition of the groups, and classified the apes according to their characteristics (height and size). He sometimes counted more than thirty individuals at a resting site. Bingham also established a list of foods that the gorillas ate, studied their vocalizations, and was charged by an adult male silverback (who was killed and brought back to the camp at Kabara). He remarked that the gorillas were primarily terrestrial, but that they could also climb trees. They were, however, less agile than chimpanzees. He noted that they avoided water, but if necessary they would cross shallow streams. Recalling, even at this period, the use of sticks as tools by gorillas in captivity, Bingham said that he never observed this type of behavior in the wild. Concerning the social organization of the primates, he noted the strong cohesion within each group. He was also interested in different ecological aspects.[16]

After the preliminary attempts by Nissen and Bingham, the picture was not greatly encouraging: studies in the field were costly, research conditions were difficult, the primates fled from the researchers, and the hours spent in the field yielded few results. The challenges in establishing systematic field research of great apes were many: funding was difficult to find for this type of study, which received little credit; logistics set in place based on existing infrastructure and with the agreement of the colonial authorities; and new methods and techniques needed to be developed, as well as protocols for verifying and evaluating the data. It was thus a new epistemological space that had to be created from the ground up, respecting the criteria that would assure the scientific integrity of the work.

Clarence Ray Carpenter, Barro Colorado, and the Howler Monkeys

From 1931 to 1933, a third student of Yerkes, Clarence Ray Carpenter, set up camp on the island of Barro Colorado in the Canal Zone of Panama.[17] He observed mantled howler monkeys at this site, which had been classified as a natural reserve since 1923. Primarily interested in vocalizations and their social function, he also studied all other aspects of the monkeys' existence: social and sexual relations, locomotion, feeding habits, mating and territorial behavior. A student of comparative psychology, he believed that the methods derived from the naturalist tradition were insufficient and unsuitable. Carpenter therefore developed different techniques to collect and analyze the data. He also used certain technological innovations, notably recording machines that allowed the playback of recorded sounds (playback technique).

Attempting to bring laboratory methods into the field, Carpenter proposed a controlled version of in situ studies. He wanted to deal with "facts" and count on the scientific method to guard against any subjectivity or abusive interpretation. His real issue was the question of the professionalization of research on primates living in their natural environment. Within a clearly functionalist framework, Carpenter proposed a sophisticated vision of primate societies. He saw these societies as being made up of networks of highly organized and complex emotional affiliations and identities. Carpenter was the first to show primates' social organization: male and female adult couples accompanied by their offspring. He also noted that grooming served not only the biological role of eliminating parasites but a social role as well, and he stressed the flexibility of behaviors. He further explained that the young learned complex behavior and acquired various social abilities during the long period of infancy. Finally, he offered a more nuanced vision of dominance, qualifying it as "relative dominance," albeit still with a dominant male having priority over the rest of the troop in access to food and sex. The principal organization of these societies thus continued to be centered around the competition between a number of aggressive males. Carpenter became a pioneer in scientific studies of primates living in their natural environment as well as a leader in primate ecology.

The Killer Ape

Primates would also play a significant part concerning a major question that would have a determinant impact upon field primatology: the origin of human societies. It was one of the principal motivations for studying primates in their natural environment. In *The Origin of Man and His Superstitions* (1920), Carveth Read described the ancestor of humans as a creature something like an ape, come down from the trees to become a terrestrial predator and hunter on the open savannas. Four years later, the Australian anatomist and paleoanthropologist Raymond Dart discovered *Australopithecus africanus.* He wondered what life might have been like for this early hominid, but he had practically no way of knowing.

Shortly thereafter, the British anatomist and paleoanthropologist Wilfrid Edward Le Gros Clark posed the question of humans' place among the primates in a major work, *Early Forerunners of Man* (1934), which focused on comparative morphology applied to both living species and fossils. Turning the Department of Anthropology at Harvard into a center for physical anthropology in the United States, Earnest Albert Hooton was also one of the first to think that the study of nonhuman primates was of primary importance in advancing research into the process of "hominization," the process of becoming human. In the early 1950s Dart took up the savanna model proposed by Read and conceived of ancient hominids as walking upright and being large-game hunters using weapons. Also supported by Lorenz, these hypotheses all centered upon a fundamental trait, aggression.

The Quest for Origins

One of Hooton's students, Sherwood Larned Washburn, a brilliant representative of the discipline, took up the idea of a "predatory transition from ape to man" set forth by Dart.[18] He wanted to revolutionize physical anthropology: "We need new ideas, new methods, new workers," he proclaimed.[19] Driven by the Western post-Darwinian obsession that combined primates and evolution, he set himself the task of reconstructing the universal story of human origins.

In an article written with Chet Lancaster in 1968, he proposed the long-lasting paradigm of prehistoric humans as hunters and killers.[20] As the master behavior pattern of the human species, hunting would have dictated a way of life, catalyzing the social, technical, and cognitive adaptive characteristics that would develop and become proper to humans. Bipedalism freed the hands and allowed for the use of tools. New types of cooperation and sociality came into being: collaboration between males, alliances linked with hunting, sharing of food, sexual division of labor, family structure.

This story of human origins was developed and popularized by the science writer Robert Ardrey during the 1960s, particularly in *African Genesis: A Personal Investigation into the Animal Origins and Nature of Man* (1961). Hominization was thus placed within a context of violence, aggression, and predation. Males were "naturally" dominant, protective, and competitive. Having to take care of the children, women were "naturally" dependent and submissive. Become carnivores, human beings developed a killer instinct, which eventually would itself lead to culture and art, as displayed in one of the opening scenes of Stanley Kubrick's film *2001: A Space Odyssey* (1968).

The New Physical Anthropology

Falling within the practice of classification (notably the classification of "races," later denounced as irrelevant and as racist science), physical anthropology had initially purported to study the biological bases of hominization through the examination of morphological and skeletal variations within the human species across time and space. Breaking free of this viewpoint, which rose from the culture of natural history collections, Washburn initiated a profound paradigm shift. Closely linking physical anthropology and primatology, he believed that the observation of different primate species in their natural environment and the exploration of the common adaptive traits between humans and other primates constituted key elements in understanding the psychological and social organization of ancient hominids. Fossils, skeletal remains, skulls, and stone tools were solid and long-lasting elements, as well as apparently being the only available vestiges, to explore the question of origins, but they no longer sufficed. Primates were thus placed in the role of "proto-humans" and became evolutionary and socioecological models. They were deemed to be so promising that people vulgarly said to be "primitive" were also brought into the question of origins.

Baboons, already used as representatives of primate societies by Zuckerman, were once again considered to be perfect models. They lived on the savanna. Their societies seemed to be centered around males that displayed aggressive behavior, acted as patriarchs, and showed a strong sense of hierarchy. Wanting to verify this baboon model, Washburn sent a number of students into the field, including Irven DeVore, who observed baboons in Kenya beginning in 1959. Study of the baboons confirmed the idea of dominance, which had become a central concept in primatology, but in a sense different from Zuckerman's. Group cohesion was assured not by a sexual logic but rather by a hierarchy based on males and established based on their degree of aggressiveness. The fact that the distribution of peanuts, meant to attract them, intensified the competition between males was totally ignored. In 1975, Sally (Slocum) Linton wrote an article called "Woman the Gatherer," denouncing the "Western male bias in anthropology" as well as a biased speculative reconstruction in which "man," "human," and "male" were equivalent. She placed the gathering of food (which accounted for the majority of food resources) chronologically before hunting, believing it was the evolutionary catalyst that led to walking upright.

With Washburn, the discipline of physical anthropology went from a paradigm that highlighted racial typology and the idea of classification to one marked by a biological perspective and evolutionary processes applied not only to humans but also to ancient hominids and those primates closest to humans. In 2005, 60 percent of field studies were tied in one way or another to either Washburn or Hooton. Adrienne Zihlman was a brilliant student of Washburn's and a specialist in comparative anatomy, which for her was not limited to bones but also included the flesh, muscles, joints, and ligaments (long neglected), as well as the relation between the parts, in a holistic vision of anatomy. Faced with the concept of a universal "Man the Hunter" imposed as the motor of human evolution during the 1960s through the 1980s, Zihlman finally attributed a role to women in the process of hominization.[21] Sociobiology (so named by Washburn and DeVore, though the term's meaning differed from the sense in which it was later used by E. O. Wilson) and its genetic fitness maximization strategies also entered into the debate. Washburn thus led primatology into a new era, open to complexity and multidisciplinary studies. In addition, the idea emerged that human societies could no longer be thought of without recourse to biology. Apes and monkeys were thus used to answer a question, "What is Man?" to which they were, evidently, completely indifferent.

Three Wise Monkeys

Japanese primatologists were also asking questions about the origin and evolution of human societies. However, influenced by their culture and their way of thinking (connected with an intense feeling of belonging to a common world), their relationship with primates was different than that of American and European researchers.[22] Their traditional belief system, Shintoism (polytheist and animist) tends toward empathy and a profound respect for all living beings. Essentially sacred, animals and plants possess a soul (*rei, tamashii*). Monkeys (*saru*) are, moreover, messengers and mediators between humans and gods. In addition, Chinese culture and Buddhism were introduced to Japan in the fifth and sixth centuries. The famous three wise monkeys ("see no evil, hear no evil, speak no evil") of the Buddhist sect Tendai are in fact Japanese macaques.

Within Shintoism and Buddhism there is no strict demarcation between humans and animals. While reincarnation in human form opens particular possibilities along the path toward illumination and awakening, compassion for all beings is required. The fact that some scientists greet their chimpanzees before work sessions or organize memorial services (*kuyou*) for them is evidence of the singularity of relations between primates and Japanese scientists.[23]

The Japanese do not deny, however, that differences exist between humans and primates. They believe, for example, that while primates may be intelligent, they cannot attain wisdom. It sometimes even happens that the macaques are seen negatively, for example, as scapegoats or as beings characterized by grotesque mimicry. There also exists a Japanese unease about a possible transformation from monkey into human and vice versa. However, they see the dissimilarities in a more nuanced fashion, rather than seeing them as unbreakable boundaries.

One of the pioneers of Japanese primatology, Masao Kawai, used the term *kyokan* to describe the method he put in place to study the macaques. Mentioned in his work *Life of Japanese Monkeys* (1969), kyokan means the feeling of living in a personal relationship marked by reciprocity and empathy. Personal ties were thus established between humans and monkeys, who shared long periods of their lives. Kawai believed that these ties in no way impaired the objectivity of the researchers. They were, rather, indispensable in getting to know better the primates being observed.

Japanese Primatology

Apart from this religious, cultural, and epistemological foundation of Japanese primatology (in Japanese, *reichourui-gaku*), the fact that two subspecies of macaque, *Macaca fuscata fuscata* and *Macaca fuscata yakui,* are found in Japan, also makes a big difference. There is in fact a single, very limited population of primates in Europe, found only on the Rock of Gibraltar: the Barbary macaque. In the United States and Canada, there are no species of endemic primates, although many species of monkeys are found in Central and South America. In Japan, however, macaques have haunted the religion, traditions, popular stories, folklore, and art for over twelve hundred years. To Americans and Europeans, on the other hand, primates are necessarily exotic, linked to long voyages of exploration or to colonization.

Toshitake Amano of Osaka University was the first scientist to undertake systematic studies of Japanese macaques, during the 1930s. A member of his laboratory, Yoshiaki Maeda, met Konrad Lorenz in 1953. Maeda introduced European ethology to Japan between 1955 and 1965. Bringing together ethology and comparative psychology, this school of primatology has a different perspective than that of Kyoto University, which is oriented more toward sociology and anthropology. Within the latter, the researchers of the Primate Research Group began field research on different groups of macaques under the guidance of Kinji Imanishi (lecturer at Kyoto University) and two of his students, Junichiro Itani and Shunzo Kawamura. They quickly noticed that there were variations between the groups in behavior and habits. They also established parallels with human societies.

Societies of Primates

An anthropologist and ecologist, Imanishi developed his own methods, inspired by those used in sociology, psychology, and anthropology. He continued an approach he had used during a study of Mongolian wild horses: individual recognition. He also carefully read the publications of Carpenter, who would visit Japan three times between 1964 and 1974. Japanese researchers considered that primates formed a "society" as an ensemble of individuals. Interested above all in these societies of monkeys (*shushakai*), they were the first to defend the idea of culture among animals. As early as the 1950s, they also asserted that macaque communities have a matriarchal organization.

Their work progressed along three main axes. First of all, they advocated the use of anthropomorphism and strategic empathy, which they justified by invoking the phylogenetic proximity between humans and primates. They then proposed a structural vision of primate societies (individuals, units, regrouping of units, communities) and were interested in the way in which each individual was a part of the group. Finally, they insisted on the importance of conducting long-term field studies, with habituation and identification of every member of the troop.

In order to do this, they decided to provide food to the monkeys. Having been chased from some regions, the macaques had become fearful of humans, which led a number of studies to fail. At Koshima, the primates had been fed by local people over a number of years, with the goal of protecting the surrounding fields. Once they were habituated to the presence of the scientists, the first task was to identify all the members of the groups being observed and to give them all names. It was indispensable to recognize each individual in order to understand the complex relations that were woven throughout the groups. Furthermore, as the Japanese insisted from the outset on long-term studies and followed the macaques for the entire duration of every individual's existence, they were able to trace their individual histories as well as that of each group.

Also, influenced by their holistic vision of the world, the researchers took into account the socioecological context in all its complexity. A specialist in Japanese primatology, Pamela J. Asquith reported a very interesting remark made by Yukimaru Sugiyama, who thought that socio-ecological and sociobiological approaches did not leave enough place for the individual: "At least in mammals including primates each has its own motivation, thought and feelings and soul in its own behavior."[24] He thus advocated an openness to complexity and explanations that included many variables, including the point of view of the animal.

Cultural Traditions Among Macaques

The Japanese macaques on the Koshima peninsula were among the first mammals to bring evidence of cultural transmission to the attention of scientists. In September 1953 the primary school teacher Satsue Mito observed for the first time a female macaque, later named Imo (meaning "sweet potato"), at the time a year and a half old, meticulously washing the sweet potatoes that Mito had given the macaques. Imo first did this in a stream near the ocean and later in the ocean itself. The young macaque seemed to have found an efficient solution for washing the sand off the potatoes while at the same time improving their taste with salt.

Carefully observing her way of doing this, Imo's brothers and sisters quickly copied her technique. Her mother was the first adult to appropriate this behavior. The adult males, however, never embraced this practice—instead they took by force the sweet potatoes already washed by the younger macaques. When she became a mother, Imo at first washed the sweet potatoes herself for her children, which then learned how to do it themselves. By 1962, nearly three-fourths of the troop were washing their food in seawater.

The technique of washing developed by Imo was more difficult to do with the grains of wheat distributed by the researchers: the monkeys had to go farther out into the water and let go of the grains, at the risk of losing them. Also, the macaques did not, in principle, know how to swim. It was again Imo who first took the risk and let go of the wheat in the water. Quickly understanding that the grains floated, she developed the habit of washing them as well. The rest of the troop soon followed suit. Bit by bit, the primates became more used to the water, eventually going in up to their waists. Some juveniles even learned how to swim. In the 1970s a new behavior emerged within these "potato-washing" troops at different sites: a kind of solitary game using small stones, a stone-handling behavior. Excepting some adult males, individuals of both sexes and all ages began making "rhythmic sounds by tambourine-like performance."[25]

Mentioned in most of the publications about animal cultures, the macaques of Koshima have left a lasting mark on the history of ethology and primatology.[26] For the first time, scientists showed evidence of invention, adoption, and diversification of a technique (later become a tradition) within a society of nonhuman primates. After more than sixty years of continuous research over nine generations, the macaques are still being observed. They continue to wash their sweet potatoes in the sea, although the behavior is slowly beginning to disappear; because they only get sweet potatoes several times a year, the occasions for learning the washing technique have become more rare.

The Kyoto University Primatological Expeditions

Toward the end of the 1950s, the Primate Research Group and the Japan Monkey Center sent numerous expeditions, including the Kyoto University Primatological Expeditions, to different parts of the world to find research sites and establish the distribution of gorillas and chimpanzees. In 1957, an expedition was organized for Southeast Asia. In February 1958, seasoned alpinist and experienced traveler Kinji Imanishi left for Africa for the first time, with Junichiro Itani. A number of their students later conducted studies at various sites in Uganda, Rwanda, Cameroon, Tanzania, and Congo, notably Masao Kawai, Hiroki Mizuhara, Yukimaru Sugiyama, Toshisada Nishida, Takayoshi Kano, and Akira Suzuki.

The Japanese researchers had a serious advantage over the other scientists doing fieldwork: having already conducted long-term studies on macaques in their natural environment, they began their work with proven methods and techniques, a sound theoretical framework, and experience in the field. They could, moreover, make comparisons with the results obtained from the studies of macaques. Upon their arrival in Africa, they thought there were cultural variations among the primate communities, whether chimpanzee or gorilla. In 1961, Imanishi, Itani, and Shigeru Azuma met Jane Goodall at Gombe. A year later, Toshisada Nishida selected a site at the foot of the Mahale Mountains in Tanzania. He began to provide a number of troops of chimpanzees with sugarcane and bananas.

It was not until 1965, however, that he officially began long-term research at Mahale Mountains National Park.[27] Beginning in March 1966, the chimpanzees came regularly to the camp, attracted by the sugarcane. Nishida gradually habituated them to his presence. He was soon in a position to study them. He tried first of all to understand how the society of chimpanzees that he had under observation functioned and evolved. The evolution of the first human communities was at the center of his research themes.

African Encounters

In 1966, Imanishi retired. Itani, a major figure in his own right in Japanese primatology, took charge of the organization of field studies. He continued to explore Africa, though he worked with smaller teams or even alone, in contrast to traditional expeditions which, characterized by large numbers of participants, offered fewer chances for encountering great apes. The Japanese showed that the chimpanzee societies were patrilineal, composed of males and females that divided into subgroups for their daily activities (fission-fusion). They also showed that female chimpanzees left their birth troop at adolescence, which avoided problems of consanguinity. They also reported on infanticide and conflicts between communities.

In addition, they documented the use of tools as well as things used for comfort or body care by chimpanzees. They collaborated, furthermore, with local populations, gathering local stories about the great apes. Later, they would be among the first to reveal the importance of grooming. This behavior of cleaning and looking for parasites allowed the management of moments of relaxation and privileged interactions between two individuals (mutual grooming) or multiple partners (social grooming). This form of "reassurance" between individuals increased the social cohesion of the group. From the very first expeditions in Africa, the Japanese were also interested in gorillas. Itani and Imanishi, for example, explored the Kigezi Gorilla Forest in 1958. Bonobos were not neglected either: Takayoshi Kano began the first long-term studies of pygmy chimpanzees in 1974 at Wamba.

The Japanese established exchanges with American and European researchers at the end of the 1950s. Imanishi and Itani met Carpenter, Coolidge, Leakey, Washburn, and Harlow in 1958. Imanishi founded the first review in the world dedicated to primatology with the support of the Japan Monkey Center and the Rockefeller Foundation (for the English version). *Primates* appeared in December 1957 in Japanese and in 1959 in English. Its editor in chief was Juichi Yamagiwa. The Japanese adapted the articles to the constraints of international scientific journals. Asquith thus explains that in addition to two major schools, there were two types of primatology in Japan: international Japanese primatology and traditional Japanese primatology.[28]

A Change of Paradigm

At the start of the twentieth century, research done in the field was still considered anecdotal. Numerous factors would, however, make the study of primates in their natural environment necessary: evolutionist ideas, the development of an international movement for the conservation of nature, the concept of an *animal-subject* having its own *world,* the growth of the discipline of ethology, the new agenda for physical anthropology, and the quest for human origins. Fieldwork was also indispensable in order to study the social organization of the primates as well as their territorial or seasonal behaviors. In addition, scientists wanted to verify the value of certain hypotheses and to complete the results obtained in the laboratory. Furthermore, researchers feared that the primates would become extinct before they had the chance to study them seriously. After World War II, the young discipline of primatology shifted the old logic of museums and natural history collections, whereby fieldwork was left to explorers and "popular primatology," to one in which work was done in the field by the scientists themselves. This change brought about multiple transformations for the primates as well as the researchers.

A chimpanzee at Tchimpounga Chimpanzee Rehabilitation Center using a wooden tool to crush nuts. The primate employed either the pointed or the flat end, according to the task.
(Photograph © Chris Herzfeld)

SEVEN

Socialities, Culture, and Traditions Among Primates

When the Boundary Between Humans and Apes Blurs

IT WAS NOT UNTIL AFTER WORLD WAR II that systematic, comprehensive, and long-term research on great apes began, led by Jane Goodall and Toshisada Nishida in Tanzania, followed by Dian Fossey in Congo and Rwanda, Biruté Galdikas in Borneo, then Takayoshi Kano in Congo. These studies underwent a tremendous development during the 1970s. Research sites multiplied. After having been seen as skulls, captive beasts, or laboratory rats, and having gone through different, often miserable, forms of existence in museums, zoos, laboratories, and medical institutes, great apes became actors in a determinant adventure for primatology. Researchers became formidable spokespersons for primates, unveiling fascinating aspects, heretofore unknown, of their lives, finally studied on their home ground. They described personalities, told life stories, detailed genealogies, reported on sophisticated technical and social abilities. Their stories were radically new.

At the end of the century, various primatologists, specialists in chimpanzees and orangutans, brought together their data from their long experiences in the field. They showed that the groups of primates studied exhibited specific cultural traits, different from one site to another, and passed down from one generation to the next in a nongenetic manner. One of the primary characteristics of human populations, *culture* (generally thought of as a form of separation from nature) was thus discovered in the natural world. Added to the evidence among great apes of a political and moral sense, aesthetic sensitivity, empathy, sophisticated social practices, and the use of tools, this discovery once again upset the rigid boundary erected between human beings and other primates.

Reciprocal Agreements

As Dian Fossey explained, any observer is an intruder (even an enemy) to wild animals. At the beginning of their research, field primatologists find themselves in the position of a hunter. They look for tracks, mark the places where the primates have passed, the sites where they feed or where they sleep. But they quickly exchange this posture of a hunter for that of a "tamer" or "shepherd," according to Hans Kummer.[1] Attempting to hide oneself while trying to study eminently social beings can only lead to establishing a climate of mistrust and truncated encounters.

Researchers thus choose to show themselves and let themselves be observed by those whom they wish to observe. Later, they gradually draw nearer, habituating some troops to their presence. With a lot of patience, they try to become accepted as a somewhat peripheral member of the group. Once they become familiar to the primates they are studying, the researchers gradually are considered inoffensive and can finally observe their subjects from a few meters away without their fleeing. The primates alone judge what is an acceptable distance.

As Biruté Galdikas said, primatology is "a science of collaboration, and not of control." It happens sometimes that the apes even invite the researchers to aid them in their daily problems, using them, for example, as protection from some predators or as a defense against local populations. The primates, at times, implicate researchers in their relations with other groups, using them as what Dominique Lestel calls a "social prosthesis."

A Science of Collaboration

Far from being passive objects of observation, apes became active subjects. They mobilized sophisticated social abilities and engaged in complex interactions with scientists. Having left their preferred space, the laboratory, researchers found themselves on the apes' territory, in places they were not familiar with, which shifted the balance in heretofore radically asymmetric relations. Primates became themselves observers of the primatologists come to observe them. This is made clear in an anecdote related by Dian Fossey:

> The group had been day-nesting and sunbathing when I contacted them, but upon my approach they nervously retreated to obscure themselves behind thick foliage. Frustrated but determined to see them better, I decided to climb a tree, not one of my better talents. The tree was particularly slithery and, try as I might, no amount of puffing, pulling, gripping, or clawing succeeded in getting me more than a few feet aboveground. Disgustedly, I was about to give up when Sanwekwe came to my aid by giving one mighty boost to my protruding rump. . . . Finally able to grab on to a conveniently placed branch, I hauled myself up into a respectful semislouch position in the tree about twenty feet from the ground. By this time I naturally assumed that the combined noises of panting, cursing, and branch-breaking made during the initial climbing attempts must have frightened the group on to the next mountain. I was amazed to look around and find that the entire group had returned and were sitting like front-row spectators at a sideshow. . . . That day's observation was a perfect example of how the gorillas' sense of curiosity could be utilized toward their habituation.[2]

In this radically different place of research, primatologists were obliged not only to change their practices, develop new methods and invent new observation techniques, but also to *transform themselves.* Immersed in the world of the great apes and wanting to be a part of it without disturbing them, some researchers went so far as to adopt forms of "becoming-animal" in the sense given by Deleuze and Guattari, by beginning to appropriate some of their behaviors: "During the early days of the study at Kabara, it was difficult to establish contacts because the gorillas were not habituated or accustomed to my presence and usually fled on seeing me. . . . Open contacts, however, slowly helped me win the animals' acceptance. This was especially true when I learned that imitation of some of their ordinary activities such as scratching and feeding or copying their contentment vocalizations tended to put the animals at ease more rapidly than if I simply looked at them through binoculars while taking notes."[3]

In the course of time spent daily with the apes, researchers experienced a growing feeling of connection with their societies, a sense of belonging to an essential community, the outlines of another way of being in the world. This experience leads to a mutation of ethos, which may explain the strength of field primatologists' engagement with those they study. These transformations also play an important role in exchanges with the primates, which are extremely attentive to the many signals given unknowingly by humans: body language, facial mimicry, smells, sounds, and so on. The fact that the researchers see the apes as full-fledged partners and approach them in a respectful manner, with a profound knowledge of their way of life, necessarily initiates a fundamental change in their point of view.

Often driven initially by the desire to experience a great "adventure," the primatologists are also transformed by the fieldwork itself. This requires a very physical engagement, a full-time commitment of body and soul, an "immersion in the rich texture of the world," in the words of Bruno Latour. Out of this long immersion in a nature both harsh and majestic also emerges an intense feeling of belonging. When the primatologists are far from their camps and forests, they often feel a vibrant nostalgia for the time spent in the field, the beauty of the landscape, the special rhythm of life that takes place in nature, the nights spent working by the light of lanterns, the intense ties to the great apes.

Travellers' Rest

The Kenyan paleoanthropologist Louis Seymour Bazett Leakey played a determinant role in the new field of research of great apes in their natural environment. He was contacted by Max-Walter Baumgärtel, who had a small hotel, the legendary Travellers' Rest, in southwestern Uganda. The site could not be better for the study of mountain gorillas: situated facing Muhabura volcano in the Virunga Mountains in front of what is now the Mgahinga Gorilla National Park, near the border between Rwanda and Uganda to the north, and Uganda and Congo to the west.

Leakey's former secretary, Rosalie Osborn, arrived at Kisoro in October 1956 and spent four months following the gorillas and noting everything that they ate. She studied their nests, vocalizations, movements, and locomotion. Jill Donisthorpe, older, more experienced, and trained in the sciences, took Osborn's place in February 1957. She worked in the field for eight months. When the Japanese researchers Kawai and Mizuhara went to study gorillas in the same region, they often referred to data collected by Donisthorpe on the gorillas' group size, the habitat (flora, fauna, and climate), their nests, diet, locomotion, and vocalizations. In spite of the time spent in the field, Donisthorpe did not manage to habituate any of the groups to her presence. She was thus unable to get close enough to the gorillas to get to know their family social organization. A number of other researchers also stayed at Baumgärtel's Travellers' Rest, including Dart, Coolidge, Imanishi, Itani, Schaller, Reynolds, Kortlandt, and Fossey.

After these initial attempts, Leakey engaged three young women, later called Leakey's Angels or the trimates (from *tri* and *primate*). Jane Goodall was asked to work on chimpanzees in Tanzania in 1960; Dian Fossey on mountain gorillas in the volcano region situated between Congo, Rwanda, and Uganda in 1963; and Biruté Galdikas on orangutans in Borneo in 1971. The three young women made a big splash in the media, particularly in *National Geographic.* A large audience thus associated their names with adventures in the field involving the great apes. A number of other studies, however, would also be organized on different species of primates during this same period, although many of them would remain unknown to the general public.

Pioneering Fieldwork

Toward the end of the 1950s, Kenneth Hall was studying baboons in South Africa. Around the same time, George B. Schaller conducted field studies of mountain gorillas for nearly two years in the Virunga Mountains. He wanted to do rigorous scientific research on their "natural" behaviors. Accompanied by Kay, his wife, and his professor John Emlen and his wife, Schaller explored the volcano region between Uganda and Congo (today the Democratic Republic of Congo) beginning in February 1959.

Schaller stayed in Africa a year longer than Emlen. He surveyed the populations and studied their area of distribution. With the invaluable aid of his African guides, he habituated certain groups to his presence and was able to approach to a distance that varied between 5 and 150 feet. He identified individuals and studied their behavior and vocalizations as well as the ecological context. He spent more than 466 hours observing the gorillas and published a detailed monograph on the species (*The Mountain Gorilla,* 1963) as well as a book for the general public (*The Year of the Gorilla,* 1964). In the latter readers found a version of gorillas that finally conformed to reality. Although his fieldwork could not be said to be a long-term study, he became the world expert on the species.

In 1964 the Catalan primatologist Sabater Pi began the first research on western lowland gorillas, in Equatorial Guinea. Hans Kummer, trained by Heini Hediger in Zurich, started studying hamadryas baboons in Ethiopia in 1961. Vernon Reynolds and Frankie Reynolds (his anthropologist wife) observed chimpanzees in Budongo Forest beginning in 1962. They stayed only eight months. In 1965 Reynolds published a monograph on the life of the species in its natural environment.

Beginning in January 1960 (and thus before Jane Goodall, who began in July of the same year), Adriaan Kortlandt also went to Africa to conduct field research. He crisscrossed the continent until 1965, from east to west, passing through Uganda, Congo, the Congo Republic, Ivory Coast, Gabon, Cameroon, and many other countries. In 1967, he was one of the first to organize experiments in the field: he placed a mechanical leopard in front of chimpanzees from the savanna and the forest in Guinea in order to compare their reactions, which he filmed. Kortlandt reported that some of the apes defended themselves by throwing sticks, which he saw as a use of weapons. Washburn (wrongly) believed this to be an incorrect interpretation: according to him, this behavior was simply part of a process of intimidation and could not be considered a use of weapons.

In 1965, Thomas T. Struhsaker showed that the vervet monkeys in the Masai-Amboseli Game Reserve (today Amboseli National Park) in Kenya used thirty-six different calls.[4] He insisted on the importance of studying the modes of primate communication within their natural environment. All these studies took place within a movement of renewal in the discipline of primatology that developed after World War II. Nevertheless, what distinguished the studies conducted by the Japanese researchers and Leakey's Angels from most of the others was the fact that they were done over a long period of time, in a systematic fashion, with a willingness to explore every aspect of the primates' existence as well as to take into account the evolutionary and socioecological dimensions.

Jane Goodall and the Chimpanzees of Gombe

In July 1960, Leakey sent Jane Goodall to the shores of Lake Tanganyika in Gombe, more than five days' travel from Nairobi. She was only twenty-six years old. Convinced of the necessity of conducting research that would not only cover the entire life span of chimpanzees but follow them over the course of several generations, they had from the outset the intention of making this a long-term study. Goodall set up camp at Gombe with her mother, Vanne, who would stay only three months, just long enough to reassure the authorities, who were concerned about a young woman living by herself in the field. Goodall was not entirely alone; her Tanzanian cook, Dominic, and his wife stayed with her.

Goodall began what would become the first field study conducted at the same site for more than half a century. Her goal was to explore the behavioral continuities between humans and great apes. She inherited the questions asked by her predecessors as well as the dominant model that was applied to all primate societies (including those of ancient hominids): the baboon model, which was based on a social hierarchy centered on males and marked by aggression and domination. In addition, Goodall wanted to study why chimpanzees lived in groups and what maintained group cohesion.

Goodall added two further specific questions upon her arrival at Gombe. Like Washburn, Leakey hoped to understand how ancient hominids had lived by studying our closest cousins. Directly tied to this, the second question was asked by the American tool manufacturer Leighton Wilkie, the project's sponsor, who was mentioned in Goodall's first article for *National Geographic.* He wanted to know if chimpanzees used tools in their natural environment. Goodall's research was thus placed from the beginning in a dual perspective: the search for human origins and the question of material culture among great apes.

She would find the latter one of the research questions of her future thesis director at Cambridge University, Robert Hinde. Hinde was one of two biologists who had published, in 1949, the first study of supposed "cultural" transmission of food-gathering techniques among titmice. Grounded in the classical European tradition of ethology, Hinde furthermore supported his student in the research methods she had adopted, with the aim of describing the behaviors, interactions, and relationships between the members of the group.

David Greybeard, Flo, Goliath, and William

Passionate and indefatigable, Goodall spent on average twelve hours per day in the field. Many months went by during which she saw hardly any chimpanzees, except at a great distance. After six months, she was finally able to see them, but at a distance still of more than three hundred feet. Her binoculars were invaluable, allowing her to observe different behaviors. The great apes gradually became accustomed to the peaceful presence of the young woman. Rather than try to hide herself, she made her presence known: she went to places frequented by the apes and waited patiently, showing them that she meant no harm.

Eighteen months after her arrival at Gombe, she was thus able to approach the apes as closely as 150 feet without interrupting their activities. During the period of habituation, one of the first tasks the young researcher gave herself was to identify a maximum number of individuals, then name them. After that, she made a list of the primates present at the site and determined the composition of groups by age, sex, filiation, and social status. Influenced by the work of the Japanese researchers, she was also greatly interested in the personalities of the chimpanzees.

Less than a year after her arrival, the famous David Greybeard (the first chimpanzee to allow Goodall to approach as close as thirty feet) visited the camp of the primatologist one April evening and fed on the fruit that grew there. Thereafter he came back each day and little by little came to trust Goodall, even accepting a banana from her hand, which she described as "a wonderful moment." From then on she brought fruit with her when she followed the primates. David Greybeard sometimes sat next to her in front of the other chimpanzees. He also continued to visit the camp, accompanied by two other males, Goliath and William.

Gradually, a number of other individuals accompanied the three males, notably the female Flo and her family. In July 1963, Flo, sexually receptive, was followed by eight males during her visits to the camp. Goodall decided to provide food for the chimpanzees in order to observe them better and get good images for *National Geographic*. In 1964 one of the females who regularly visited the camp gave birth. This allowed Goodall the chance to observe the development of the infant in detail.

Fishing Sticks and Tool Use

Finally able to get as close as three feet to the chimpanzees, Goodall was in a position to report on heretofore unobserved behaviors, one of the most remarkable being the use of tools. David Greybeard was the first chimpanzee that she observed "fishing" for insects with a long stick.[5] Köhler and Yerkes had previously described the use of various tools by chimpanzees, but that was in captivity. The same kind of observation in the field was at the time very rare.[6] Goodall precisely documented these behaviors and accounted for the natural context in which they took place. She added that some apes used leaf sponges in order to get at rainwater trapped in tree cavities. She also reported that the chimpanzees performed a ritual, lasting between fifteen and thirty minutes, that she called "the rain dance." When it began to rain, several males ran about, striking the ground and trees in their way, tearing off and dragging branches, all the while calling out loudly.

As Goodall's work advanced, her observations of the chimpanzees dethroned Washburn's "baboon" model. For prehistorians and anthropologists, the use of tools was proof of humanity as much as the differences in skulls. Chimpanzees manufactured and used tools much more frequently than baboons. Also, the "chimpanzee model" seemed pertinent in regard to hunting. Goodall reported that chimpanzees were occasional hunters, having spied David Greybeard tracking a young bush pig. She later witnessed many episodes of hunting. The chimpanzees tracked different prey, for example, young bushbucks or red colobus monkeys. Moreover, they offered a much more peaceful image of human origins than baboons.

War and Peace Among Great Apes

By the end of the 1970s, however, this Edenic image of chimpanzees living peacefully in a harmonious world was shattered. Goodall witnessed the elimination of one of the groups she had habituated by another. She did not know how to interpret these first observations of violence: aggression, violent infanticide, cannibalism, territorial wars. Nishida reported similar events at Mahale.

In a seminal monograph based on twenty-five years of fieldwork on chimpanzees, published in 1986, Goodall preferred to emphasize the personalities and intelligence of the chimpanzees, the sophistication of their societies, their technical abilities, and the complexity of their strategies of moving from one food source to another.[7] She showed that primates were capable of making decisions, were able to plan out short-term actions, and could categorize, classify, or put objects around them into relation, for example, estimating weight and length. Goodall called these abilities "pre-mathematical." She also attributed a degree of self-consciousness to the chimpanzees, as well as imagination and the ability to conceptualize.

With her reliable and original data gained from systematic and long-term study, Goodall revolutionized fieldwork in primatology. She did nothing less than redefine the scientificity of research in the field by focusing on the diachronic aspect of studies pursued over life spans and across generations. With Goodall, we were no longer in the presence of "apes," but with Flo, Fifi, Flint, and David Greybeard, individuals that became as familiar to the public as the ape "stars" such as Cheeta, King Kong, or Curious George.

Chimpanzees and Field Researchers

Various researchers did work at Gombe, including Geza Teleki, who published a work on predation behavior in chimpanzees in 1973. Conducting observations since 1972, Anne Pusey of Duke University supervised the electronic archives for data collected at the site over more than fifty years. Richard Wrangham of Harvard University (like Goodall, a student of Hinde's), known for a major publication (with Dale Peterson), *Demonic Males: Apes and the Origins of Human Violence* (1996), also conducted research at Gombe, between September 1970 and January 1972. Following in the footsteps of researchers on human evolution and "primate cultures," he later set up a long-term project at Kibale National Park in Uganda.

By the end of the 1960s, chimpanzees were being studied at more than thirty-five sites. After having habituated a group of around fifty chimpanzees in the Budongo Forest, Uganda, in 1966, Yukimaru Sugiyama became interested in their social interactions. He found evidence of symbolic communication, the use of objects for comfort, greeting, begging, and food-sharing behavior. He later went to work at Bossou in Guinea, increasing the number of publications on the chimpanzees of the region beginning in the late 1970s. Around the same time, William McGrew (Cambridge University) began the longest study of savanna chimpanzees at Mount Assirik (Niokola-Koba National Park, Senegal). A specialist in cultural primatology, chimpanzee technologies, and social learning, he was one of the authors of a major article on culture among chimpanzees. He often collaborated with Linda F. Marchant, who also went to Gombe and Mahale. Together they explored different questions tied to culture.

Cultured Apes

Supported by Hans Kummer, Christophe Boesch and Hedwige Boesch-Achermann began research in Ivory Coast in 1979. They have worked together for more than thirty-five years on the chimpanzees at Taï National Park, where the habitat differs greatly from that of Gombe or Mahale. Having visited these two sites, Boesch was aware from the outset of questions about differences in culture. It was only after eight years that the Boesches were able to observe the chimpanzees capturing termites using fishing sticks.

Furthermore, it had long been thought that chimpanzees in western Africa did not hunt. However, after nine years, the couple was able to show that the chimpanzees at Taï had one of the highest hunting frequencies of all known populations. The fact that a behavior has not been observed does not mean that the behavior does not exist. Christophe and Hedwige Boesch also studied the practice of breaking open nuts using stone hammers and natural anvils. These tools have such a great importance for the chimpanzees that they fight over them and bring them along when they go from one group to another.

The two primatologists also showed that mothers pass down to their children the necessary technical skills for breaking open nuts through *active teaching*, which is extremely rare. Great apes normally learn by intently observing the techniques and skills that interest them (peering). They then imitate the observed actions, practice them intensely, and finish by appropriating for themselves the target skills.

Researchers at the Department of Primatology at the Max Planck Institute for Evolutionary Anthropology (in Leipzig), directed by Boesch, are working on the three species of African great apes and three of the four subspecies of chimpanzee at many sites and in collaboration with approximately thirty-four research centers gathered under the Pan African Program. They are particularly interested in the influence of ecological conditions on the variations observed in tool usage.

During the 1990s, Vernon Reynolds resumed studies in Budongo Forest. He published a monograph in 2005 (*The Chimpanzees of the Bu-*

As part of the Pan African Programme, Christophe Boesch and his team observed a new tool-use behavior in Bakoun Classified Forest (Guinea). The chimpanzees routinely fish for green algae with long wooden sticks and twigs, sometimes up to four meters in length.
(Max Planck Institute for Evolutionary Anthropology, Pan African Programme: The Cultured Chimpanzee)

dongo Forest: Ecology, Behaviour, and Conservation) in which he confirmed that in this population of apes there were few examples of tool use. Insisting on the dimension of relations between humans and apes, he dedicated a part of the work to the question of conservation, describing the threats the chimpanzees faced. Finally, a number of studies (including those of Wrangham and Nishida) were done on the pharmacopeia of the chimpanzees, which seem to have knowledge of some forty plants used to treat different illnesses. They have shown that most species of great apes eat plants in order to feel better. This "self-medication" speaks to different hypotheses that have sprung up following the theories tied to the "Man the Hunter" model. The process of hominization may not have begun as a movement from the forest to the savanna, nor from an economy linked to hunting, but from a mixture of gradual anatomical and behavioral innovations, driven by selection and tightly connected to the complex interactions established within the forest environment (compare the primate-angiosperm coevolution theory, the arboreal theory of primate evolution, and others).

Dian Fossey: Eighteen Years Among the Mountain Gorillas

Arriving in Africa at the end of December 1966, Dian Fossey set up her first camp at Kabara (at ten thousand feet altitude) on the site formerly occupied by Akeley and Schaller in the former Belgian Congo. She was thirty-five years old. Financed by *National Geographic* and the Wilkie Foundation (as was Jane Goodall), Fossey was accompanied by Alan Root to help guide her initial steps. She was also aided by three African assistants, including her faithful Sanwekwe. Beginning her epic experience without any kind of training, Fossey had only Schaller's book and several articles, including the two published by Goodall in *National Geographic,* as guides.

Before being able to do this type of reporting for the American magazine, however, she faced many unexpected challenges. At the beginning, the gorillas of Mount Mikeno were very aggressive. Fossey was attacked many times by the males, which terrorized her. Very serious accidents often happen in the field, and attacks or kidnappings (as was the case at Gombe and Kabara) can occur, arising from instability in the surrounding regions, though they are not widely publicized. After several months, the gorillas finally came to accept Fossey's presence, and she began to accumulate hours of observation. In July 1967, she was forced to interrupt her research when she was assaulted and taken by Congolese soldiers.

Giants of Rwanda

Seeking refuge in Rwanda, Fossey founded the Karisoke Research Center, between Mounts Karisimbi, Mikeno, and Visoke, on September 24, 1967. Very close to Karisoke, Volcanoes National Park, a Rwandan extension of Albert National Park, stretches over nearly fifty square miles and shelters thirty-two groups of gorillas (six of which have been habituated to humans). Installing her "gorilla observation camp" on Mount Visoke (at ten thousand feet), Fossey closely followed four groups out of nine whose members she regularly observed.

In 1970 she left for England to work on her doctoral thesis in zoology with Robert Hinde. She received her doctorate in 1974 from Darwin College at Cambridge. Titled "The Behaviour of the Mountain Gorilla," her thesis was considered one of the most complete reports ever written on a primate species observed in its natural environment. In her first publication in *National Geographic* ("Making Friends with Mountain Gorillas") in January 1970, Fossey gave readers a completely different image of gorillas than they had been accustomed to. They read about Rafiki, Samson, Peanuts, and Geezer grooming each other and playing games. They met a number of families, including that of Uncle Bert. They were moved by Coco and Pucker, baby gorillas rescued by Fossey after their family had been slaughtered.

In 1983 she published her famous *Gorillas in the Mist,* in which she wrote of her observations gathered over thirteen years among the giants of Rwanda. By the end of the book, readers had come to know the personalities of a number of individuals, followed from infancy to adulthood. We saw them grow up, followed them in their daily activities, and sometimes were present at their death. As only someone who had shared their existence over many years could do, Fossey described, simply and with great empathy, the apes she loved. She showed the role of the male silverbacks, the most experienced in the troop, explaining that they assured the defense and cohesion of the groups. She also analyzed the movement of sexually mature females from one family to another, and pointed out the fact that dominant males sometimes called upon one of their sons to help defend the group. She noted that the females formed alliances to protect themselves from males.

Fossey furthermore explained the degree to which fights between families could be ritualized in order to avoid serious conflict, once the individuals had gained enough experience. She depicted the relations between mother and infant as well as the different stages of development of the young. Fossey completed this picture by providing details about the different aspects of the gorillas' existence: the construction of their daytime and nighttime nests, their vocalizations, their instinctive distrust of unknown objects, the solicitude of certain individuals toward older or weaker members of the group, their grooming, their diet. Assassinated on December 26, 1985, Fossey is buried in the cemetery she had created for the gorillas, close to the tomb of her favorite, Digit, killed by poachers on December 31, 1977.

Gorillas and Scientists

Ian Michael Redmond had very close ties with Dian Fossey, centered around their shared passion for gorillas. With the killing of Digit (to which both Fossey and Redmond were very close), he added the role of conservationist to his work as a field biologist. He has studied gorillas as well as other species such as elephants for more than thirty years. Temporarily closed due to the Rwandan genocide, the Karisoke Center was set up at Ruhengeri (today Musanze) and pursued its mission under the auspices of the Dian Fossey Gorilla Fund International.

Jordi Sabater Pi (University of Barcelona) was also interested in the three species of African great apes and worked on questions of culture and conservation. He had the privilege of observing Snowflake, a rare albino western lowland gorilla of the Barcelona Zoo who died in 2003. Purchased by the Catalan primatologist, Snowflake had been captured in 1966 in the Rio Muni region (western Cameroon), where Sabater Pi had been doing research. Working in the field for more than thirty years, Juichi Yamagiwa (Kyoto University) is one of the rare researchers to have studied western lowland gorillas at Moukalaba-Doudou (Gabon) and eastern lowland gorillas at Kahuzi-Biega National Park (or Grauer's gorillas, *Gorilla beringei graueri*), as well as the mountain gorillas of Virunga (*Gorilla beringei beringei*), both in Democratic Republic of Congo (DRC). Toward the end of the 1980s, Liz Williamson studied western lowland gorillas at the Lopé Reserve in Gabon as well as eastern lowland gorillas at Kahuzi-Biega National Park (DRC).

After working for a number of years on the three species of African great apes, Annette Lanjouw was named director of the International Gorilla Conservation Program, a post she held for fifteen years. She is today the vice president of the Great Apes Program at the Arcus Foundation, an important donor for conservation, field projects, and American primate sanctuaries. Magdalena Bermejo (University of Barcelona) left for Africa in 1986 and has become an authority on western lowland gorillas after having studied them at the Lossi Gorilla Sanctuary and Odzala National Park (both in DRC). Although they had had little previous contact with humans, the Lossi gorillas allowed Bermejo to conduct long hours of direct observation within the group. She is also very active in their conservation as well as in the fight against the Ebola virus.

Since the 2000s, many families of western lowland gorillas have been part of a program of habituation funded by the World Wide Fund for Nature (WWF) in the Dzanga-Sangha National Park (Central African Republic). In January 2016, twins, which are quite rare, were observed at Bai Hokou in the Munye group, one of the first groups to be habituated at the site (1998). Classified as critically endangered by the International Union for the Conservation of Nature (IUCN), western gorillas include two subspecies, western lowland gorillas (*Gorilla gorilla gorilla*) and Cross River gorillas (*Gorilla gorilla dielhi).* The latter live in a region situated between Nigeria and Cameroon and are currently the most threatened subspecies of great ape, with a population of only 250 to 300 individuals.

Conservation and Ecotourism

Dian Fossey categorically rejected the idea of ecotourism although, paradoxically, the revenue generated by tourism became the strongest motive for conservation for the governments of the three countries involved.[8] According to Schaller, there were 450 mountain gorillas in 1960. Fossey counted 275 between 1971 and 1973, this number going down to 254 individuals in 1981. Today there are more than 700 individuals. Some people even set the number at 880 gorillas for the two groups, those of the Virungas and the Bwindi Impenetrable National Park. They are the only great apes whose numbers have increased since the 1980s. The courage and dedication of the rangers of Virunga National Park, such as the veteran Innocent Mburanumwe, cannot be stressed enough. The chief warden, Emmanuel de Mérode, was nearly killed in April 2014. More than 140 rangers have died defending the park from poachers and militia groups.

WWF supports conservation programs for eastern gorillas, which include mountain gorillas and Grauer's gorillas, the largest subspecies of gorilla and the world's largest primate, which is now listed as critically endangered by the IUCN. These two subspecies are extremely threatened by trafficking, bushmeat hunting, the exploitation of natural resources in the region, continuing armed conflicts, and the displacement of refugees. The population of Grauer's gorillas has gone from approximately 16,900 in 1995 to around 3,800 as of 2016.

A Woman in the Indonesian Forest: Biruté Galdikas

Born at Wiesbaden (West Germany) in 1946, of Lithuanian parents who sought refuge in the United States after the Second World War, Biruté Marija Filomena Galdikas studied anthropology at the University of California at Los Angeles. One of her professors mentioned Jane Goodall's research during a course. Galdikas was only nineteen years old, but she knew the cause to which she would devote her life. She met Leakey in 1969 during a conference at the university, and founded Camp Leakey in the heart of Tanjung Putting National Park in November 1971. She was the first Westerner to set up camp in this natural reserve, where the trafficking of baby orangutans was common. Local people as well as officials often adopted them as pets and objects of prestige.

It was in the middle of this tropical forest that Galdikas learned of the death of Louis Leakey on October 1, 1972. Nonetheless, she pursued her research. From the time her center opened, she combined rehabilitation with the study of wild orangutans. She established that their life span was approximately fifty years and estimated the interval between births to be eight years (the longest among the great apes), a factor that reinforced the vulnerability of the species. She showed that they ate sap, leaves, sprouts, and stems as well as more than four hundred species of fruits and flowers. They also ate bark and insects, such as termites. In *Reflections of Eden: My Years with the Orangutans of Borneo* (1995), Galdikas declared that they were "the finest botanists in the world." Having a very precise memory of the topography of their territory, the orangutans regulated their movements according to the places and periods of fruit becoming ripe. They were forced to develop sophisticated abilities for reaching available food, which was often both sparse and difficult to access. They protected themselves, for example, with gloves made from plants to manipulate spines and showed a great ability in peeling fruit.

The orangutans also displayed what could be called a veritable mechanical genius, demonstrated in their building of nests (Galdikas called them "works of architecture"). They also emitted a great variety of sounds: grunts, long calls (very long and powerful, made by males), and fast calls (a rapid succession of identical sounds, also made by males), as well as the vocalizations and yelps of the females and the cries of the young. Galdikas also described the use of large leaves as umbrellas and the construction of shelters in bad weather. A male named Henry was seen crossing a river by walking on logs.

As for the social organization of the species, people speak of "semi-solitary" primates. The dissemination of resources and the small quantities available do not allow many individuals to fulfill their needs in the same area. Carel Philippus van Schaik (University of Zurich), one of the greatest specialists of the species, says that some adult males rule over large territories in certain regions of Borneo, limiting the movements of other individuals. Endowed with a formidable social memory, orangutans maintain important relations between individuals over time and in different forms (as sexual partners, within traveling bands, or in temporary groups feeding, for example, at the same fruit tree).

Observing Orangutans

During the 1970s, other studies were also being done on the Asian great apes. Richard K. Davenport undertook an eleven-month study at Sabah (in the Malaysian part of Borneo), but he gathered very little information as the orangutans were very difficult to observe in this location. Around the same time, John MacKinnon and Peter Rodman were also studying orangutans in Borneo, at the Kutai Reserve.[9] Two major research centers existed on the island. Ketambe Station (Gunung Leuser National Park, Sumatra), oriented more toward research, was created in 1971 by Herman D. Rijksen with Indonesian scientists. The other, oriented more toward rehabilitation, was founded at Bohorok in 1973 by the Swiss zoologists Regina Frey and Monica Boerner with the aid of the Frankfurt Zoological Society and WWF. This center would later be administered by the Indonesian authorities. David A. Horr was also among the first scientists to study the apes of Borneo.

In 1993, Carel van Schaik reported that the orangutans made instruments designed to get honey or trap insects at Suaq Balimbing (less than fifty miles from Ketambe on Sumatra), where he set up a monitoring camp. The English scientist Ian Singleton worked there from 1996 to 1998. Later, he would become the director of the Sumatran Orangutan Conservation Programme, whose center took in orangutans confiscated by the authorities and reintroduced them into the forest. The camp at Suaq Balimbing was abandoned in 1999 because of political troubles. Van Schaik and Singleton rebuilt it in 2005.

Van Schaik would report numerous other instances of tool use by orangutans during studies done at Gunung Leuser National Park (North Sumatra) and Gunung Palung National Park (Borneo). He went to the University of Zurich in 2004. The biological anthropologist Cheryl D. Knott (Boston University), who regularly collaborated and published with van Schaik, was interested in the influence of the environment on the orangutan's way of life as a means to better understand the evolution of human societies. Van Schaik coauthored a pioneering article on cultural variability among orangutans in 2003.[10] One year later, he published a monograph on the species in which he talked about the emergence of culture and tradition in orangutans.[11]

Anne Russon, a professor of psychology at York University (Toronto), has conducted research on orangutans since 1989. She has published a number of articles with Galdikas, especially on the question of imitation. She continues to work in the field at Kutai National Park (East Borneo). The primatologist Isabelle Lackman-Ancrenaz and the veterinarian Marc Ancrenaz have studied orangutans in Sabah Forest in Malaysian Borneo for a number of years. They established the Kinabtangan Orang-utan Conservation Project in 1998, the year in which they met Jenny (then mother of an infant), the first orangutan they habituated. The Ancrenazes also work on problems of conservation as well as conflict resolution between local populations and the primates. Akira Suzuki does research in Borneo with the Gunung Palung Orangutan Project in Kutai National Park Kalimantan. Many primatologists participate in conservation organizations, for example, the Orangutan Land Trust led by Michelle Desilets. Camp Leakey, started by Galdikas, is still active in Tanjung Putting National Park.

Bimaturism

This fieldwork greatly increased researchers' knowledge of orangutans. In 1978, Rijksen showed that male orangutans were the only great apes that experienced two phases of maturity (bimaturism). They attain sexual maturity during the first stage (unflanged males), although their morphological differences, of both size and weight, with females (sexual dimorphism) are much less pronounced than during the second stage (flanged males). Mating occurs, but the males in the first stage of development use different reproductive strategies than those in the second. The latter have priority over the females.

At the first stage, the subadults are between eight and fifteen years old; at the second (connected with a high level of testosterone), the adults are between fifteen and twenty. They develop a "facial disk" that makes their appearance rather spectacular (once large males lose their dominant position, their cheek pads, or flanges, recede). Their muscle mass increases, as do the long hairs on their arms and back. The throat swells and they get a double chin, since their laryngeal sacs enlarge, allowing the males to amplify their long calls, which play a role in territorial defense as well as sexual selection. These signature vocalizations mark territorial limits and thus help avoid conflicts. They are also made to communicate with receptive females seeking the protection of these big males. Van Schaik further explained that these long calls have a secondary function of delaying the second-stage development of nearby males by triggering hormonal stress. Furthermore, he found, with Gauri Pradhan (University of South Florida), that male Sumatran orangutans have themselves the capability of delaying the onset of the second stage of maturity. They thus prolong their adolescence, becoming bigger and stronger. They are then able to rival the big males that monopolize the females in their area.[12]

Kano and the Bonobos of Wamba

Having studied chimpanzees, gorillas, and orangutans, primatologists turned toward the last described species of great apes, bonobos. Up until the 1980s, this species was almost unknown to the general public and was still little studied in the 1990s. Bonobos live exclusively in the Democratic Republic of Congo. The first field research was conducted at four sites: Wamba (Kano and Kuroda), Lomako (Badrian and Badrian), Tumba (Horn and Nishida), and Yolisidi (Kuroda and Kano). These studies began in 1972 with the American anthropologist Arthur Horn. However, the bonobos in the region near Lake Tumba were so timid that in the space of two years he observed them for barely six hours.

Takayoshi Kano of Kyoto University had better luck. Kano left for the field in 1964 with Imanishi, Itani, and many of their students. With the help of Nishida, already set up at Mahale, Kano continued the studies already begun in the Lake Tumba region in February 1972. After a year and a half of research, he located the bonobos on the west bank of the lake, one of the hydrographical limits of the species' distribution, which also include the Congo River, Lake Ndombe, and the Lomami, Kasaï, and Sankuru Rivers. These natural barriers have long isolated the bonobos, placing them in a situation of insularity.

On November 1, 1973, he encountered his first pygmy chimpanzee, then discovered the site where he would conduct systematic research on bonobos beginning in 1974, Wamba (the Luo Scientific Reserve). Kano provided food for five or six groups that lived near villages. The study done at Wamba, the first long-term one done of the species, is recognized as the gold standard on bonobos. It continues to this day, although it was interrupted during the armed conflicts in the DRC (1991–94, 1996–2002). Many bonobos were killed. Current estimates suggest that a minimum of fifteen thousand to twenty thousand individuals live in the four known strongholds of the species. The total number of the population is difficult to know.

Sexual Practices Among the Pygmy Chimpanzees

The distribution of bonobos covers an area with a stable environment that is very rich in resources. Chimpanzees spend on average 25 percent more time than bonobos looking for food. Taking advantage of their free time, it seems that bonobos have been able to develop refined socialities marked by empathy, certain forms of sociosexual "ritualization," playful activities, and a very rich sexuality. Brian Hare, Victoria Wobber, and Richard Wrangham explain these behaviors, and even certain psychological as well as physical and morphological traits (for example, the juvenilized craniums characteristic of the species), by the "self-domestication hypothesis." Bonobo characteristics may be due to selection against aggression: less aggressive and juvenile behaviors, the sharing of food, and a greater tolerance would have been selected, while aggressive actions and offensive attitudes would have been eliminated.[13]

As opposed to bonobos in captivity, "wild" bonobos have rarely been observed using tools during foraging activities. Hunting varies from group to group. Researchers think that bonobos hunt prey such as forest antelopes, squirrels, or other rodents, but at some sites, for example, at Lilungu, Martin Surbeck and Gottfried Hohmann have seen evidence of the hunting of colobus monkeys. However, they reported that at Wamba the two species have established "friendly" relations, going so far as to groom one another.

Takayoshi Kano proposed the idea that the study of wild bonobos could be a key to explaining how the first human societies functioned other than through aggression and violence. The primates did not disappoint him. They excel in the peaceful management of their societies. Adult unrelated females establish intense intimate relations (for example genito-genital rubbing) that play an important role in the formation of cooperative female alliances. Mothers remain close to their sons, who stay in their natal group. The fact that they are related prevents aggressive behavior by other males. Furthermore, certain sexual rituals allow the easing of tensions that arise within groups, reinforcing the ties between members. Among bonobos, sexuality is not reduced to its reproductive role, but also assumes a social function.

Kano published *The Last Ape: Pygmy Chimpanzee Behavior and Ecology* in 1992. He has amassed thousands of hours of direct observation, to which must be added those of Takeshi Furuichi, Akio Mori, Genichi Itani, and other Japanese researchers who have also spent many hours in the field. In 1982 Suehisa Kuroda (University of Shiga Prefecture) had already published a book on bonobos: *Pygmy Chimpanzee: The Unknown Ape.* Beginning in 1989 Kuroda worked in the Nouabalé-Ndoki Forest (DRC), which has populations of both gorillas and chimpanzees. He was particularly interested in the use of medicinal plants among western lowland gorillas and the chimpanzees of the region, comparing them to the bonobos at Wamba.

A Colonization of *Pan paniscus*

In 1973 Noel and Alison Badrian (State University of New York at Stony Brook) arrived in DRC with few means and armed only with their courage. They created a site at Lomako near Befale (the Lomako Forest Pygmy Chimpanzee Project). They confirmed a great number of the observations made by the Japanese researchers, even though the environment was different from that of Wamba, particularly in terms of resources. Randall L. Susman (SUNY, Stony Brook) arrived at Lomako in 1979 as coordinator, followed by Nancy Thompson-Handler, Richard Malenky, and Frances White. In 1984 Susman edited the first English-language monograph on the species: *The Pygmy Chimpanzee: Evolutionary Biology and Behavior.* Gottfried Hohmann (Max Planck Institute) joined the team in 1990 and began to work with Barbara Fruth at Isamondje. The civil war in 1991 brought an end to most of their work. The social, economic, and political situation was catastrophic. The massive displacements of populations, the trade in bushmeat, deforesting, and burning constituted grave threats to the primate populations.

Only two research sites still functioned in 1992: Wamba, where data has been continually collected for more than twenty years, and Lomako, where Barbara Fruth and Gottfried Hohmann observed bonobos from 1990 to 1998. The young Belgian primatologists Jef Dupain and Ellen Van Krunkelsven, supported by the Royal Zoological Society of Antwerp, arrived in 1994 at Isamondje, in the camp newly set up by Fruth and Hohmann, for training. They were supervised by Linda Van Elsacker who, since 1992, had studied the group of bonobos at the Planckendael Zoo at Malines near Antwerp. The university and the Antwerp Zoo were privileged centers of the study of primates: they were located in the port city where all the specimens, living or dead, arrived from the former Belgian Congo.

Although still a student, Jo A. Myers Thompson set up camp at Yasa, a region bordered to the south by the Sankuru River, in 1992. She founded the Lukuru Wildlife Research Project at the southernmost site of the bonobos' distribution. She observed pygmy chimpanzees that used both forest and savanna, which opened up new perspectives in the field of research on the species, particularly regarding their behavioral flexibility. In order to help protect the primates, she created the Bososandja Faunal Reserve, home to six hundred individuals. Thompson courageously continued to travel to the DRC during the conflicts. Additionally, a Spanish team, composed among others of Sabater Pi and Bermejo, worked at Ikela in the Lilungu region in 1993. Kortlandt also explored areas in the 1990s that might be homes for populations of pygmy chimpanzees.

In 1995, Dupain and Van Krunkelsven created their own camp at Iyema (near Lomako) and conducted studies for a year. However, the perilous situation in the country as well as the economic and political chaos tied to the seizure of power by Laurent-Désiré Kabila in 1997 made fieldwork impossible. The Spanish researchers who worked at Lilungu and the German and American primatologists abandoned their camps. The scientific reserve at Luo and Salonga Park was evacuated.

Forests Emptied of Their Animals

In 2000 several ecological studies were started up again. But the assassination of Kabila in January 2001 followed by the rise to power of his son Joseph began a new wave of troubles. Later, those who returned to the forest after the first calm periods often had the impression that it had really been "emptied." Today, observations continue to be conducted at a number of sites, including the long-term studies at Wamba (where the center is currently directed by Takeshi Furuichi of Kyoto University) and Lomako. Having stayed in 1983 at Lomako, then being forced to leave in 1998 because of the war, Frances White (University of Oregon) continued her research in 2005. The Max Planck Institute also conducted fieldwork that began in LuiKotale in 2002 within Salonga National Park. Fruth worked there, particularly focusing on self-medication among bonobos in collaboration with Van Elsacker and Dupain as part of the LuiKotale Bonobo Project, founded in 2005.

Still director of the Lukuru Wildlife Research Project, Jo Thompson is also involved in the Tshuapa-Lomami-Lualaba Conservation Landscape Project, directed by the conservationist Therese Hart.[14] Sheltering some ten thousand bonobos, this area for pygmy chimpanzees had remained nearly unexplored up until 2007. Finally, Brian Hare (Duke University) established comparisons between humans, chimpanzees, and bonobos (3chimps—Hominoid Psychology Research Group) in collaboration with the Max Planck Institute. Rather than working with captive animals, he did his experiments (respectful and noninvasive) in three African sanctuaries, testing problem-solving abilities with the bonobos from the Lola ya Bonobos sanctuary in the DRC (the only sanctuary in Africa to take in this species, since 2005), and with the chimpanzees at the Tchimpounga Chimpanzee Rehabilitation Center (Republic of the Congo) and at the Ngamba Island Chimpanzee Sanctuary (Uganda).

Cultures and Traditions Among Great Apes

The project director for the Firestone Plantation Company of Liberia, Harry A. Beatty, observed in 1951 a chimpanzee breaking open a palm nut set on a stone with a rock. In the early 1960s Goodall drew the attention of the scientific community as well as the general public to the fabrication and use of tools among the great apes in their natural environment. A number of articles and books addressing this question followed, documenting different cases of tool use. In 1980 Benjamin B. Beck published a detailed study of the use of tools in the animal world.[15] Nearly half of the reported cases involved nonhuman primates. While aware of the risk of "chimpocentrism," Beck reported that chimpanzees led all other tool-using animals in both frequency and variety.

In 1999, a further step was taken. Recognized authorities in primatology who had worked for many years in the field, as well as specialists on the question of material culture among primates, published a joint article on their findings.[16] They described the use of tools, weapons, measuring sticks, and objects used during social interactions or games, as well as the use of instruments for comfort and care, such as leaf sponges, twigs for cleaning noses, leaf brushes, flyswatters, sticks for masturbation, and so on. They showed the existence of certain social rituals (such as handshakes and above-the-head hand holding during grooming) and other rituals, such as one tied to the arrival of rain.

Technical know-how, instruments and tools, and rituals and habits differ from one troop and one site to another. Transmitted nongenetically from generation to generation by social learning, they constitute, for the primatologists, forms of "culture." The first author, Andrew Whiten (University of Saint Andrews) is a specialist in social learning, culture, and tradition in primates. One of the other authors, William McGrew, is a specialist on tool use among chimpanzees and the evolutionary dimension of primate cultures.

Four years later, the existence of "cultural traditions" was also shown among orangutans. These primates have a heightened sense of comfort: they use handfuls of leaves to clean their face and body, sticks for scratching themselves, bundles of greenery for pillows, leafy branches as fans against insects, gloves made from plants to handle spiny fruit, sticks to masturbate, and protective cushions to sit on ground covered with spines. They construct roofs above their nests as protection from bad weather or as sun shades.

Rare Tool Behaviors

Observations of the use of tools or other instruments are extremely rare among gorillas. Thomas Breuer (Max Planck Institute) reported the case of a female named Leah living in Nouabalé-Ndoki National Park (Republic of Congo). She was seen using a branch both as a walking stick and as a measuring stick to judge the depth of water and stability of the soil. At the same site, a female belonging to another family, Efi, was observed breaking a piece of wood from a trunk. She then placed it on the swampy ground and walked on it upright, using it as a bridge to get across the muddy surface. Only a single case of tool use has been shown among mountain gorillas, even though numerous systematic studies have been done on this subspecies since 1959. A study was done in 2013 on one of the eight habituated groups at Volcanoes National Park (Rwanda), the Kwitonda group. After having watched her father put his hand in a hole in the ground to catch ants, a juvenile female, Lisanga, chose a piece of wood to use to fish for the insects, enabling her to reach them without getting bitten.[17]

Researchers have noted the rarity of tool use associated with the search for food among bonobos. Kano has described, however, the use of leafy branches as protection against rain. He also reported the case of the exploration of a termite mound with a stick. Bonobos, like chimpanzees, use instruments for social exchanges, play, comfort (flyswatters, scratching sticks), bodily care (toothpicks, "towels" made from plants), and protection (leaf umbrellas, rain hats). The context is often playful and a greater variety of tool use is seen among females than males. At Lomako, observers have noted the use of leaf sponges designed to collect water, even when water is directly accessible elsewhere. The chimpanzees of Bossou also make these kinds of sponges. They even sometimes drink alcohol (from naturally occurring fermented palm sap). The drinking of this alcoholic nectar, tied to their fruit-based diet, brings various benefits, being rich in vitamins and minerals as well as glucose. Some individuals show signs of drunkenness.

Elsewhere, various populations of chimpanzees living in the Goualougo Triangle (Congo) as well as in a region situated between Nigeria and Cameroon use rarely described tools, for example, beehive pounders

Set of tools used by the adult chimpanzee Farafa to hunt bush babies that take refuge in tree cavities, Fongoli, Senegal, June 2015.
(Photograph courtesy of Jill Pruetz)

(tools used to break open beehives to get honey), ant-digging sticks, food pounders, and grating stones. At Tongo (Virunga, DRC), chimpanzees dig in the ground to get at tubers during the dry season. The biological anthropologist Jill D. Pruetz (Iowa State University) has reported the use of "spears" by chimpanzees of Fongoli in southeast Senegal.[18] Ian Tattersall notes that this is the only case of regular hunting with a "weapon" among nonhumans.[19]

Living on the savanna (in the hottest and driest of all the sites where great apes are studied), the chimpanzees of Fongoli (males, females, juveniles, and infants) use pointed branches that they thrust into tree cavities where bush babies (*Galago senegalensis*) hide themselves during the day. Pruetz has observed how the chimpanzees make these instruments of predation. They first strip the leaves from long branches, then sharpen the point with their teeth. The chimpanzees at Fongoli are also distinguished by their use of caves and their surprising attitude in the face of brush fires, which they observe calmly, without fear, anticipating its movements and effects.[20]

A Primatology Set in Stone

Despite the variety of instruments used by great apes, primatologists who study their cultural traditions focus on the use of tools, particularly stone tools. Breaking open nuts using a stone hammer upon a natural anvil is thus highlighted. The use of a "chopper" to open a special fruit, *Trequila africana,* has also been often described. These percussive technologies are considered the most complex instrumental practices in the animal world as well as the closest to early human stone technologies. This makes sense within the paradigm whereby the parallel between apes and ancient hominids is used to clarify the emergence of human culture. Durable and long lasting, stone tools constitute some of the rare remains that can attest to human progression, through the series of tools carefully reconstructed by prehistorians.

The notion of "tool" is thus nearly always linked to "stone." A new example of the use of stones by chimpanzees (the subspecies *Pan troglo-*

dytes verus) occurs at different sites in western Africa. In what at first may appear to be strange "rituals," chimpanzees throw stones at trees or hit them against the trunk while making pant-hoot vocalizations (which often characterize combative displays). Are these ritualized behavioral displays in which the sound is amplified by the stones striking the trees? Does the practice itself lead to its becoming a sort of ritual, disconnected from the original intent of intimidation? In spite of the possible use of stones as ritual elements, they are nonetheless regarded as tools.

Instruments made of fibers, on the other hand, ephemeral and made for comfort or body care, are described but less valorized. Fiber, a soft material, is also traditionally seen as more feminine, part of "nature," whereas stone is hard, masculine, part of "culture." Fibers (implicitly associated with needlework and sewing) belong in this way to the private sphere; stone (tied to the control of nature) to the public sphere. In reality, the fact that most of the tools and instruments used by primates are made of fibers has long been neglected. These fibers come in a multitude of forms in the forest: trees, creepers, bark, leaves, herbs, vines, twigs, sticks, branches, mosses, fruit, and so on. In the works by Beck and Whiten et al. cited above, out of thirty-eight tools used by chimpanzees, only five are made of stone (of which two or three are made of both stone and fiber). Among orangutans, not a single stone tool is found out of the nine listed. Of the total of sixty-five behaviors and instruments noted among chimpanzees, fifty-eight are linked to fibers. The proportion is even greater among orangutans: out of thirty-six instruments, thirty-four are of fibrous materials.

The segregation between stone and fiber, hard and soft, masculine and feminine, culture and nature, exploitation and comfort is summed up in the statement of a delegate to a workers' congress in 1867, reported and commented on by Michelle Perrot: "'For men, wood and metal. For women, family and cloth.' A great material and symbolic division of the world. The hard for men; the soft for women."[21] Today, researchers are more attentive to these uses of plants and fibers, the different forms and functions tied to the same tool, and the context within which the tools are used. Nold Egenter proposed the idea of a "soft Prehistory" and advocated a greater attention to fibro-constructive material cultures. Primates are both fiber users and practitioners of fibro-constructive techniques as well as tool users and tool makers, abilities that they combine in extremely creative ways.

Mothers and Technicians

The paradigm in which we think of ancient hominids has long been haunted by the idea that technical skill and tool use belong in the domain of typically masculine abilities. Jane Goodall, however, stated that Gremlin, a female, was one of the best specialists in her community at capturing insects by using branches and other sticks. Furthermore, she passed on this aptitude to her children. Christophe and Hedwige Boesch added that females do better than males in this kind of activity, especially in breaking open nuts. Elizabeth Lonsdorf noted that adult female chimpanzees had better scores than males during tests for tool use at all levels (frequency, duration, efficiency). Four years of field research at Gombe allowed her to point out certain differences in the development of this type of technical know-how, for example, in fishing for termites: young females were twenty-seven months ahead of young males, an advantage they maintained. As for the chimpanzees of Fongoli, Jill Pruetz has shown that more females than males hunt bush babies with sticks used as spears.

How to explain that primate mothers are better technicians than males? Different factors come into play. The daily life of females is, first of all, more complicated than that of males: they have to perform all of the same everyday activities as males do, but in addition they also take care of the young. They are obliged to assume responsibility for all of their offsprings' needs for around four or five years. The primatologist and mathematician Jeanne Altmann sums up this situation with the phrase "double career." The result is that, through necessity, mothers develop an ability to do two things well at the same time. This also means they get more practice than the males in all the techniques they use daily.

Mothers break open nuts for their young, for example, as well as themselves, so they perform the same series of actions more often, which naturally results in improved manual dexterity and increased expertise. Also, their curiosity is maintained by the fact that they are constantly surrounded by the young, who are much more curious and exploratory than adults. Consequently, females are often at the origin of technical innovations. In Japan, the female macaque Imo is a famous example. Furthermore, males can use force to obtain some things that they then do not have to bother to get for themselves.

Female Primates as Custodians of Culture

Young primates live for a number of years with their mothers, who pass down all of the indispensable knowledge the young need to make a life for themselves in the forest: a knowledge of fruits, plants, and fibers as well as the places and periods of fruiting; knowledge of medicinal plants; how to move through the trees; what behaviors to adopt vis-à-vis predators; a cartographic representation of their territory, and so on.[22] Mothers serve as models for learning techniques such as nest building and the fabrication and use of different tools and instruments. They also guide their young through the tangle of social status, alliances, and hazardous relations at the heart of their troop. The large number of individuals in the same group, which can number over a hundred among chimpanzees, makes the management of relations between individuals even more complex.

This handing down of knowledge, techniques, and ethos from the most experienced to the youngest is made possible by specific characteristics shared with humans, particularly an extended infancy, neotenic characters shared by all the great apes, and a long life span. This transmission from mother to infant constitutes a decisive element in evolution. Integrating into a new troop after their sexual maturity (among chimpanzees, bonobos, and gorillas), females also bring habits, traditions, and traits from one group to another. They comprise the driving forces of the cultural dynamics at work among communities of great apes. Females are thus not only formidable *transmitters* of culture but also remarkable *diffusers* of it.

Living with the Great Apes

Primates were made to play the role of proto-humans in order to shed light on the process of hominization. The "baboon model" that had long captured the scientific imagination gradually unraveled. Jane Goodall proposed that chimpanzees were good models for understanding the evolution of ancient hominids. In contrast to baboons, they seemed to form peaceful and harmonious societies. This gentler vision of our origins was, however, challenged by an unexpected dimension shown by the chimpanzees of Gombe in the wars, infanticide, and acts of cannibalism observed by Goodall. Zihlman then proposed his "pygmy chimpanzee hypothesis," which made bonobos the better prototype of a common ancestor for the human species.

Primates nonetheless partly escaped the role of static models for thinking about human origins that primatologists wanted to assign them by showing their exceptional behavioral flexibility and by the extraordinary variability of their societies. They forced researchers to look at them differently and to ask them other questions. Female primatologists also began to challenge the traditional roles attributed to male and female primates as well as views of their social organizations as principally centered around males as leaders, defenders, and police. This time, it was females who emerged transformed by a vision that distinguished itself from a "masculine" orientation and patriarchal bias. In accordance with the holistic vision advocated by Washburn, primatology became more open to complexity, whether of individuals or groups, of sophisticated socialities, or of primates' environment and its different components.

In this change of paradigm, thanks to systematic, long-term field studies, great apes became "actors as authors" observed within their own world.[23] Both males and females showed themselves to be subtle strategists, highly developed socially, experienced technicians, skilled tool makers and fiber users. They were also endowed with individual personalities and a particular life history punctuated by infancy, adolescence, midlife crisis, and menopause or andropause. Primates in these communities constructed and tested their social skills through affiliations, competitions, struggles for power, alliances, cooperation, reassurances, or reconciliation. They learned and experimented with technical innovations and new ways of doing things. Families and troops became places for the transmission and strengthening of abilities (social, cognitive, and technical), different forms of knowledge (in botany, self-medication, and gathering and hunting techniques), habits and practices, and even "rituals." There emerged a way of living, style, ethos, culture, and specific traditions unique to each troop.

Researchers finally attributed to primates capacities that had long been denied them, initially because of what Dominique Lestel terms the "hygienic split" between humans and animals. Utterly changing the relations between researchers and observed subjects, work done in the field would also transform the collective representations of great apes and provoke a radical change in the way we look at those beings that many now accept as our closest cousins.

Jane Goodall observing and being observed by the female chimpanzee Fifi in Gombe, Tanzania, 1962.
(Hugo Van Lawick/National Geographic Creative)

EIGHT

Women and Apes

Sex, Gender, and Primatology

DRESSED IN SHORTS AND A FIELD SHIRT, her hair in a ponytail, a young blonde woman is kneeling at the foot of some large trees in the Gombe Forest. She holds out her hand toward a very young chimpanzee who watches the woman, both curious and cautious, then stretches out its own hand until their fingertips touch. This image has become emblematic of the first long-term field research done by women on the great apes, in the 1960s, when Jane Goodall came to the attention of the general public in *National Geographic* magazine. The first article she wrote about her experience, "My Life Among Wild Chimpanzees," appeared in August 1963. Readers discovered a courageous explorer alone in the jungle, following and observing chimpanzees in the field, crouched down with a notebook on her knee. They read about her washing her hair by the riverside, drinking a cup of tea in her tent accompanied by the chimpanzees she had come to know, and watching the sunset over Lake Tanganyika.

This type of publicity placed women at the center of field research on primates. Before this time, however, women had received very little recognition in the fields of science, medicine, or anthropology, even though many of them excelled despite the difficulties they encountered breaking into these heavily male-dominated domains. Women's liberation movements were fighting decisive battles to take their place in patriarchal societies. In the scientific world, however, position, financing, and honors had until this time all been reserved for men. Jane Goodall became an iconic figure, the grande dame of primatology. Arriving in Africa several years after Goodall, Dian Fossey also became a star, portrayed by Sigourney Weaver in *Gorillas in the Mist* (1988). Based on Fossey's book of the same title published in 1983, the film introduced a large audience to her adventures, made all the more stirring by the impressiveness of the gorillas and Fossey's tragic assassination in 1985.

With Goodall and Fossey, women researchers were finally made visible, even valorized. Their lives in highly exotic locations in Africa fascinated young women, who dreamed of a life in the wild and searched for independence at a time when primatology itself was developing into a distinct scientific discipline. Supported by the articles in *National Geographic,* primate research became known to a larger public, presented largely as a feminine endeavor through coverage of researchers such as Goodall, Fossey, Biruté Galdikas, and Shirley Strum (the so-called *National Geographic* effect).

In reality, at the time these primatologists were conducting their fieldwork in the 1970s, women made up slightly less than 20 percent of the membership of professional primatological societies. In the 1980s women accounted for 30 percent of the membership of the American Society of Primatologists; it was not until 1992 that they surpassed men at 52 percent.[1] In 1999 Alison Jolly wrote that there were the same number of women as men in the professional primatological societies.[2] Nevertheless, the general public believed that women were much more numerous than men in the different disciplines associated with primate study.[3] In Japan, a nation known for its expertise in the field, the majority of researchers were men, as were nearly all of the major figures of the Japanese schools of primatology. The impact of the publications of the first women field researchers and the attention they received in the media gave rise to the idea that they ruled the discipline, even though they rarely occupied the most important positions. This chapter explores the scientific and metaphorical links that have existed between women and great apes since the end of the nineteenth century.

Early Women Explorers

Women explored the world long before they became interested in primates. Some left for the field as early as the end of the seventeenth century. Nuns or spies, disguised as men or dressed in crinoline, women crossed deserts, navigated rivers, collected butterflies, and went in search of distant peoples. In 1699, the German naturalist and entomologist Anna Maria Sybilla Merian, at the age of fifty-two, organized the first objective scientific expedition, to Suriname in South America, accompanied by her daughter. In the nineteenth century, Ida Pfeiffer planned to travel around the world. She was the first Western woman to explore the forests of Borneo, with her Maltese guides. She would be followed in 1875 by Marianne North, who depicted the flora of Borneo and Java in her paintings. Toward the middle of the century, the Dutch aristocrat Alexandrine Tinne, the Englishwoman Florence von Sass-Baker, the American Mary French-Sheldon, and the English ethnographer Mary Henrietta Kingsley all explored the "dark continent."[4]

Western women thus had the opportunity to encounter great apes in their natural environment beginning in the nineteenth century. Having read the works of Livingstone, Du Chaillu, and Burton, Mary Kingsley developed a passion for Africa. In 1893 she left England for Angola, then traveled up the Congo River. During a second voyage, she went to Sierra Leone and down through French Congo (today Republic of Congo). She climbed Mount Cameroon and navigated the Ogowe River in Gabon, all the while collecting specimens. Upon her return she gave a number of lectures and wrote of her adventures in two books, *Travels in West Africa* (1897) and *West African Studies* (1899). During the same period, Margaret Fountaine studied and collected butterflies on five continents. In her travels she explored both East and West Africa. Women adventurers even explored the antipodes. In 1882 Daisy May Dwyer (known as Daisy Bates, the surname of her second husband, John Bates) traveled to Australia and Tasmania. At the beginning of the twentieth century, she decided to live among the Aborigines in Western Australia. A self-taught anthropologist and indigenous rights activist, she devoted more than forty years of her life to the study and preservation of aboriginal culture.

In Love with Africa

Often traveling alone, sometimes accompanied by a spouse or traveling companion, these early explorers were also supported by their editors, who realized there was much to be gained from publishing their adventures. Their travelogues were very popular at this time of triumphant colonialism. The public was fascinated by the stories, which allowed them an escape from their daily routines and led them to discover exotic countries full of mystery and excitement. In this respect, Osa Johnson certainly numbers among those who sparked the imagination of Americans during the period preceding the Second World War. With her husband, Martin Johnson, she brought back innumerable images (as photographs or films) as well as stories from previously unexplored places. Martin and Osa Johnson brought the world, from Africa to Asia to the Pacific islands, into American living rooms.

Osa Helen Leighty's remarkable journey began in Chanute, a small town in Kansas, where she was born in 1894, and led to a wild escape with her husband from Nalint cannibals in the jungles of Malekula in the New Hebrides (today Vanuatu) in 1917. Later photographs show her posing in khaki shorts and pith helmet astride a zebra. A formidable hunter who never lost her composure, she was Martin's invaluable associate. She was responsible for the hundreds of porters employed during their African expeditions. One of the first women ever to see gorillas and chimpanzees in the wild, she was also the first to fly over Kilimanjaro, in a plane piloted by her husband. They also traveled in Borneo for two years, where they encountered orangutans, thus becoming among the fortunate few to have seen the three species of great apes known at the time in their natural environment.

Between the two world wars, women continued to travel. Emily Hahn, for example, had the project of crossing Africa from the east to the west. From St. Louis, Missouri, Emily (known as Mickey) was the first American woman to receive a degree as a mining engineer, graduating from the University of Wisconsin at Madison in 1926. She went to the Belgian Congo in 1930, living there for two years with her Pygmy guides. Ill prepared and nearly without equipment, but led by an excellent guide, she traveled over eight hundred miles on foot across Central Africa toward Lake Kivu. After spending time in Asia, she became a writer, eventually specializing in animal subjects. She had a small monkey that sat on her shoulder. In her well-known book *Eve and the Apes* (1988), she wrote about nine women closely linked with primates.

Thus, despite the manifold difficulties and particular constraints tied to their sex, at a time when there was systematic and structural discrimination against them, several strong women nevertheless succeeded in exploring unknown lands, blazing a trail for others equally passionate to follow.

Delia Akeley

Following these great women explorers, several women with strong personalities took part in various scientific and ethnographic expeditions at the start of the twentieth century. This was the case of the two successive wives of the famous taxidermist and naturalist Carl Akeley, Delia Julia Denning and Mary Lenore Jobe. During these African expeditions, these women were not content to play minor roles. Delia, also known as Mickie, traveled to the Africa for the first time in 1905, to British East Africa, where she shot the largest of the elephants in the "Fighting Bulls" group, afterward exhibited in the main hall of the Field Museum of Natural History in Chicago. During the next expedition (1909–11), during which the Akeleys encountered Theodore Roosevelt, who was taking part in the Smithsonian-Roosevelt African Expedition (1909–10), Delia killed one of the eight elephants that would later be used in the African Hall of the American Museum of Natural History in New York as one of the habitat groups Akeley proposed to the president of the museum, Henry Fairfield Osborn, during the 1910s.

Delia, however, was not simply a good shot; she was a passionate naturalist and an accomplished expedition director. From 1921 to 1922 she again lent her talents to her spouse, taking part in the Akeley African Expedition to the Belgian Congo. It was during this expedition that she saved him from death after an elephant attack, bringing him back to camp with the help of porters, and later taking care of him during a bout of blackwater fever. The principal objectives of this expedition were to furnish specimens for Akeley's dioramas as well as to establish the truth about gorillas, whose reputation had heretofore been a mix of different stereotypes: ferocity, bestial sexuality, hybrid beings between humans and primates, representatives of the missing link.

Other women also accompanied Akeley on this voyage: his secretary, Martha Miller Bliven; Mary Hastings Bradley and Alice Bradley (age five), the wife and daughter of Akeley's friend Herbert Edwin Bradley, a lawyer and naturalist from Chicago; and Priscilla Hall, Alice's nurse.[5] During this first experience in Africa, the Bradley couple, Mary and Herbert, collected a number of specimens and wrote several articles together.[6] Mary also described their adventures in numerous travel books. One of these works was mentioned by two Japanese researchers, pioneers in field studies of primates, Kawai and Mizuhara, in 1959.[7]

On her return to the United States in 1922, Delia brought back a "souvenir" from the expedition: a pet monkey named J. T. Junior.[8] The following year, Delia and Carl divorced, but that did not stop her from

Delia Denning Akeley with J. T. Jr., the little vervet monkey she adopted during an expedition in Kenya. (From Delia Akeley, "Notes on African Monkeys")

organizing further expeditions to Africa, including one for the Brooklyn Museum. During the 1930s, although more than sixty years old, she lived for several months with the Pygmies of the Ituri Forest (which has among the most amazing diversity of primate species in the world) in the former Belgian Congo, where she discovered several new species. Delia shared her souvenirs and memories, gleaned from her travels throughout East and West Africa, at conferences and lectures, as well as various articles and books, most notably *Jungle Portraits,* which appeared in 1930. She married a businessman, Warren D. Howe, in 1939, and died in 1970 at the age of one hundred.

Mary Jobe Akeley

In 1924 Carl Akeley married Mary Lenore Jobe, a graduate of Columbia University and herself already a recognized explorer. In 1926 they left for Africa. The Akeley-Eastman-Pomeroy African Hall Expedition (1926–27) had as its goals completion of the habitat groups in the African Hall at the American Museum of Natural History, documenting the landscape and botanical environment for various dioramas, and conducting a study of the recently created Albert National Park with a view to improving its management. Mary took on multiple roles for the expedition. She became Akeley's secretary, field assistant, photographer, and safari manager.

After the death of her husband on November 18, 1926, on Mount Mikeno near Kabara (today in the DRC), Mary stayed at Kabara to oversee Akeley's burial. She was among the first women to observe the gorillas of Albert Park in their natural environment. She knew their area of distribution, their attitudes and postures, vocalizations, and diet, as well as the surrounding fauna, including some that were threats, such as leopards. She recognized their tracks, observed their nests, felt their presence near her tent, noted their behaviors, and attempted to understand their psychology.

Upon her return to the United States, Mary wrote several articles about her adventures with Akeley. Named secretary of the American Division of the Comité Belge pour la Protection de la Nature (later to become part of the International Union for the Conservation of Nature), she worked to promote cooperation between Belgium and the United States to assure the proper management of Albert Park, develop other natural reserves, and place them under the auspices of an international federation. She also worked toward obtaining the necessary funding for the completion of the African Hall, supervising its construction as an "assistant adviser." She returned to Carl's gravesite in 1947, more than twenty years after his death, and wrote touchingly about their voyage in her book *Congo Eden,* published in 1950.

Other couples, such as Harold C. Bingham and his wife Ethel Lucille Forest, experienced this type of scientific expedition. Among the first scientists sent by Yerkes to Africa, Harold and Ethel observed mountain gorillas in Albert National Park for two months in the summer of 1929. Later, a number of other field primatologists worked as couples. The names and the work of the women, however, were not always as well known as those of their partner.

Creative and Collaborative Couples

Carl, Delia, and Mary Akeley are good examples of "collaborative" or "creative" couples that various researchers have described, including Georgina Montgomery.[9] If one looks closely, it becomes apparent that, moved by the same passion, a number of scientists studying primates work as couples, whether in the field, in the laboratory, or at research stations. Some women accompany their spouses on expeditions. Others support them during their university careers: they care for the young primates being studied by their husbands, take on the role of research assistants, or conduct their own work in parallel with that of their spouse.[10] Donna Haraway states that "the prominence of married couples publishing together in the primate literature is striking."[11]

In Asia, Céline Boutan took care of the young female gibbon Pépée as she would her own daughter, starting in 1907. Very close to her husband, the scientist Louis Boutan, she helped with experiments designed to understand Pépée's behavior and cognitive abilities. When Eugen Teuber, only twenty-three years old, became the director of the first primate research center in Europe in 1912, his wife Rose followed him to Tenerife in the Canary Islands. She collaborated with him, organizing the station and taking care of the chimpanzees. At the end of December 1913 it was another couple that succeeded them, Wolfgang Köhler and his wife Thekla Achenbach.

In the United States, the father of primatology, Robert Mearns Yerkes, co-wrote a seminal book of over six hundred pages with his wife Ada Watterson: *The Great Apes: A Study of Anthropoid Life* (1929). Ada received her PhD from Harvard, where she taught botany. She quickly became an essential collaborator for Yerkes, involved in both research and publications. She also took care of the researchers and students staying at their summer home in New Hampshire, ensuring that everyone had a pleasant stay that mixed work and play. Later on, she would manage the gardens and infrastructure of the laboratories at New Haven and Orange Park while continuing her own research. For example, Ada conducted experiments with Gua (the young female chimpanzee raised for nine months by the Kelloggs in 1931) when she was brought back to the laboratory.

In 1911, the Russian biologist Nadia Nikolaevna Ladygina married Alexander Fiodorovich Kohts, the founder and director of the Darwin Museum (created in 1907 in Moscow). Nadia worked in the Department of Zoopsychology, raising the young chimpanzee Joni from 1913 to 1916. She became one of the dominant figures in primatology, while her husband, with her help, tried to make the Darwin into a great museum. An admirer of Nadia Kohts and a major figure in comparative psychology, Harry Harlow also worked with his wife, Margaret Kuenne, a scientist specializing in the field of childhood development, beginning in 1948. The couple would publish many important articles on the question of affection and the creation of ties among animals, particularly primates, in the early 1960s.

Although they were themselves scientists, many wives of researchers thus played the role of the "good woman" behind their husband. They were not always mentioned in official publications, and many of them received no recognition. In a broader sense, the almost systematic way in which the contributions of women in science were barely recognized impacted the financing of their work. The men were the ones who had careers, got funding, occupied center stage, and allowed, eventually, their spouses to do research. Rarely mentioned in the history of science, many women who have remained anonymous nonetheless made important and decisive contributions to the development of the new field of research of primatology.[12]

Married, with Apes

Creative and collaborative couples are also well represented in the world of the apes that lived with human families as part of an experiment. Some candidates for learning human language were introduced into families, with the women playing the roles of substitute mother, instructor and, eventually, research assistant. Married to Winthrop Kellogg in 1920, Luella Dorothy Agger raised the young chimpanzee Gua (who arrived at the age of seven and a half months) along with their son Donald (ten months old at the time), from June 1931 to March 1932. She also worked with Winthrop to record the data. Together, they wrote the book that recounted this exceptional experience, *The Ape and the Child,* published in 1933.

Based at the Yerkes Laboratories of Primate Biology at Orange Park, Catherine and Keith Hayes, married in 1940, adopted Viki as if she were their daughter in September 1947. Catherine graduated from the University of Wisconsin in 1942, the same year as Keith, published a book about her experiences with Viki, *The Ape in Our House* (1951), and began work on another, *The Ape in Our Past,* which unfortunately was never finished. She also co-wrote a number of scientific articles with her spouse. She and Keith were later divorced, after which Catherine formed another collaborative pairing with a well-known primatologist, Henry W. Nissen, who had been among the first researchers sent into the field by Robert Yerkes. The couple lived at Orange Park, married from 1954 to 1956, with Henry conducting research at the Yerkes Primate Laboratory, where he eventually became director. They too were divorced, and Catherine then pursued her scientific career at the University of Chicago Medical School for thirty years.

The psychologists Allen Gardner and his wife Beatrix (née Tugendhat) began a program of study of the learning of human symbolic language by great apes in 1966. After having worked in the group of Nikolaas Tinbergen, the Nobel Prize–winning ethologist, at Oxford University (where she received her DPhil in 1959), Beatrix took care of several chimpanzees: Washoe, Mojo, Pili, Tatu, and Dar. The couple co-authored numerous articles and participated in many collective works. Together with T. E. Van Cantfort, they published a major work on the question of symbolic language learning by great apes, *Teaching Sign Language to Chimpanzees.*

The psychologist Roger Fouts took care of Washoe, as well as Mojo, Tatu, and Dar, after the Gardners. Throughout his career, he would benefit greatly from the help of his wife Deborah, herself a psychologist. She cared for the chimpanzees and co-wrote a number of articles with her husband. Deborah also co-directed the Chimpanzee and Human Communication Institute, created with her husband, as a fitting place of retirement for the talking apes they had known. Ann James Premack, who graduated with a BS from the University of Minnesota in 1951, also figures among the women who, in collaboration with their spouse, found a place in the history of primatology. Married to David Premack, Ann was the instructor of the female chimpanzee Sarah. She established an inventory of studies of great apes in the book *Why Chimps Can Read,* and coauthored with her husband the seminal work *The Mind of an Ape.*

These women, and many others as well, played an important role in research on primates. The collaborations they formed with their scientist husbands were creative and fruitful. However, as I have already stated, their contributions were not always recognized for their true value.

Women and Talking Apes

In the small world of studies on the language capacities of great apes, many women, however, escaped these effects and became quite well known. They were filmed, photographed, and interviewed, often in the presence of their protégés, particularly Francine Patterson and the gorilla Koko, Lyn Miles and the orangutan Chantek, and Sue Savage-Rumbaugh and the bonobo Kanzi. They began to receive media attention after the initial success garnered by Jane Goodall and Dian Fossey in the 1960s. These women benefited from the fame attained by the first women field primatologists, public interest in great apes, the popularity of the study of language, and the rise of the discipline of primatology. This time, it was the women who made magazine covers while their companions remained in the shadows.

One of them, however, received much less attention in the media than other researchers associated with language learning among great apes, although the history she shared with the chimpanzee Lucy is fascinating. Janis Carter, a graduate student in psychology at the University of Oklahoma, was given the delicate task of looking after Lucy at a crucial period of her existence, when she went from the pampered life she had always known in the United States to her new life as a "wild" chimpanzee in an African forest.

Once she reached adolescence and became much stronger than her adoptive parents, Lucy posed enormous problems: she was difficult to control, destructive, sexually restless, unpredictable, and potentially dangerous. The Temerlins wanted to get back to a normal life, but they did not want to see their "unusual daughter" end her days in a zoo—or worse, a laboratory. They wanted rather to offer her the chance to finally "live the life of a chimpanzee." They enlisted the aid of a student, Janis Carter (who met Lucy for the first time in 1976), and found a place for her in Africa: the Chimpanzee Rehabilitation Trust in the Abuko Nature Reserve, directed by Stella Brewer, the daughter of Edward Frank Brewer, who was himself the first director of the Wildlife Department of Gambia.[13]

Lucy and another talking ape, Marianne, left for Africa in September 1977, accompanied by Maurice and Jane Temerlin (another example of a collaborative couple). Janis Carter was also part of the group. At first she thought she would stay in Gambia for several weeks—then several months. In fact, she remained for many years and continued to work on conservation projects in Africa, notably as the director of the Chimpanzee Rehabilitation Trust. The Temerlins soon returned to the United States. Lucy and Marianne were placed in a small group of chimpanzees on Bamboo Island, a large, narrow island in the Gambia River more than four hours east of Banjul. Practically isolated from human contact, Carter lived on this island with the small community of apes, where she had some extraordinary experiences.

Becoming Chimpanzee

The task assigned to Janis was arduous. She had to transform Lucy into a "wild" chimpanzee, a state she had never known, having been raised as a consumer perfectly adapted to American civilization of the 1960s and 1970s. Nonetheless, "her human guardians decided that she should be a chimp."[14] In order to teach Lucy the rudiments of living in the forest, Carter first had to learn herself how to survive in the jungle. She carefully observed the actions of other chimpanzees, noting, for example, the techniques used by the most accomplished male of the small group, Dash, when he caught ants. She then tried to imitate them. Her apprenticeship extended to many daily activities, particularly choosing and gathering food, identification of potential dangers (such as crocodiles), the use of different vocalizations, and the appropriation of various techniques used in the forest.

This process of learning, her daily social environment (composed exclusively of great apes), her close personal ties to Lucy, and her isolation from human society led Janis to develop a singular "ethos," a form of "becoming-chimpanzee." She even said that it was getting more and more difficult to "become human" again. This intimate experience of what it means to "be" an ape mirrored, in a fascinating manner, Lucy's situation; she was having the same experience, but in the opposite sense. Having undergone a "becoming-human" (that had profoundly structured her personality, her sensory perception, and her way of being in the world), it was very difficult for her to "become chimpanzee" again. Although she was of an age to have young, she never coupled with another chimpanzee.

Thanks to Carter's efforts, Lucy finally accepted living among the other chimpanzees, after long having refused to feed herself and after many years of "depression." Although she never became entirely accustomed to her new life, she nonetheless became one of the authority figures of the group. Carter was eventually able to leave the island. She returned to visit her protégé on many occasions. In 1987, Lucy's skeleton was found at Carter's old campsite, with her hands and feet missing. The exact cause of her death was never determined.

Satsue Mito and the Macaques of Koshima Island

The lack of recognition women received in the scientific domain was not limited to the United States. Even in Japan, the name Satsue Mito was not very well known. However, she was responsible for the majority of data collected over decades concerning the Japanese macaques that washed sweet potatoes. A schoolteacher, Mito lived in the village of Ichiki, across from the Koshima peninsula, home of a troop of wild macaques. Her father had fed them sweet potatoes since the 1920s, gradually habituating the macaques to his presence.

Capable of recognizing each member of the troop, Mito was the first person to notice the behavior of the young female macaque Imo, the initiator of the practice of washing the sweet potatoes in the sea, which was then learned by many of the others. Day after day, Mito took notes, patiently amassed data, and took innumerable photographs, documenting this exceptional case of cultural transmission. Mito observed the macaques for over forty years, and thus carried out perhaps the longest continual collection of field data. It was not until twenty years after the start of her research that she entered Kyoto University and worked with other researchers, from 1970 to 1984. Mito's daughter, Mariko Hiraiwa-Hasegawa, is among the rare Japanese women primatologists to have gained international recognition.

Why Women?

Forming a collaborative couple himself with Mary Leakey, one of the few women paleoanthropologists at the time, Louis Leakey had long sought out collaborators capable of remaining at the same site over a period of years for long-term studies of the great apes' way of life. In 1946, Leakey made an initial attempt with a male researcher, but he lacked patience, found the time too long, and did not deal well with the solitude. The attempt failed. Leakey then decided to recruit a woman, Rosalie Osborn, whom he had met at the Museum of Natural History in London. After having initially worked for Leakey as a secretary, Osborn left to observe the gorillas on Mount Muhavura in Uganda, the region where George Schaller would later do his research. Begun in October 1956, the experience was conclusive. The young woman showed herself to be more competent than the male candidate. Only the lack of financing brought an end to her study. The following year, Jill Donisthorpe left for the tropical forest. She gathered a vast amount of data on gorillas over a period of eight months. The third woman to venture into the universe of the great apes (this time with chimpanzees) would revolutionize the world of primatology: Jane Goodall, followed shortly by Dian Fossey and Birutė Galdikas.

For Leakey, women possessed crucial qualities for observing great apes in their own environment. He believed that they were more devoted and passionate than their male counterparts while being just as intelligent. They were also more careful observers. Furthermore, habituating a group of primates to a human presence posed fewer problems for a woman than for a man because the dynamic lacked the rivalries that could exist between male primatologists and single male apes.

Goodall and Fossey were also chosen because they had not followed a strict scientific training, which may have greatly influenced their observations. Goodall had been a secretary and Fossey an occupational therapist, although she had tried, unsuccessfully, veterinary studies. Beginning her research in 1960, Goodall did not receive her PhD in ethology until 1965. Fossey began work in the Congo in 1967; she did not get her doctorate until 1974. The first candidates chosen by Leakey thus left for the field endowed with an openness of spirit and a new way of looking at things. Their lack of scientific training, however, made it more difficult for Leakey to obtain funding.

Women in the Jungle, Men at the University

Lurking behind the motivations that guided Leakey was probably the division that traditionally placed women on the side of the "natural" and men on the side of the "cultural." Linked to maternal care, the supposed proximity between nature and women conferred upon them another characteristic essential for Leakey: the capacity to communicate with individuals devoid of language. "Naturally" closer to children because of their particular role in reproduction and the fact that they are charged with deciphering the needs of newborns, women would have developed specific abilities in the domain of nonverbal communication, abilities indispensable for understanding what the apes transmit through body language or gesture.

To all this was added the problem of the availability of researchers to remain in the field over extended periods of time. Leakey wanted to put in place long-term studies. At this time, male researchers spent little time in the field because of their professional ambitions: careers were made above all in museums and at universities. Ambitious scientists could not afford to spend years in the field. Women, however, did not have the same constraints. Academia did not offer them a lot of opportunities. Men from the upper classes of society occupied the positions of influence and were granted most of the promotions and important posts. With little hope of obtaining a good post, women sought out new areas that were not already dominated by men.

Leakey hoped, furthermore, that the chosen candidates would become emotionally involved with the primates they were going to live among, that they would feel responsible for them and thus stay longer in the field. In this he was not mistaken: Jane Goodall spent fifteen years at Gombe, the center where research has been nearly continual for more than fifty years now. Dian Fossey lived for eighteen years among the mountain gorillas. Biruté Galdikas stayed in Indonesia for twenty-nine years. It must be noted, however, that a number of male primatologists, for example, Toshisada Nishida, Takayoshi Kano, and Christophe Boesch, also worked for many years in the field.

Furthermore, the engagement of the primatologists (whether women or men) was also sustained by the "call of the wild," the extraordinary sensation of freedom and their passion for the great apes. They experienced an immense liberty in conducting their research. They were freed from problems of institutional hierarchy as well as constant control over their work. Both women and men worked patiently for years, creating strong ties with the primates. Moreover, knowledge of the life of great apes in their natural environment was extremely limited at the time. Everyone therefore had the feeling of working for an important cause.

Pioneering Studies

It is important to note that well before the field studies done by Goodall and Fossey in the 1960s captured the media's attention, women scientists had been interested in the primates. They had not been waiting for Louis Leakey to come along. Despite their lesser numbers and lack of recognition for their work, several individuals put their mark on the study of primates beginning in the nineteenth century. The first authorized translator of Darwin's *Origin of Species* into French (1862), Clémence Royer was also the author of a number of works in various disciplines (physics, biology, anthropology, and economy, among others), including a study on the mental faculties and "social instincts" of primates.[15] She also commented on Victor Meunier's book *Les singes domestiques* in an 1887 article.[16] Marie Goldsmith, a doctor of science at the zoology laboratory at the Sorbonne, published a number of articles on the great apes during the 1920s, in which she studied their development in comparison to humans. She also studied the brains of gorillas and chimpanzees, showing how closely the brain of a newborn chimpanzee resembled that of a newborn human.

A biologist and anatomist, a professor in the Department of Obstetrics and Gynecology at the Yale School of Medicine, Gertrude L. Van Wagenen arrived at Yale in 1932 and founded, on her own initiative, one of the first breeding colonies of rhesus macaques in the country. Begun with the purchase of a macaque from a pet shop in 1931, this colony grew to eighty individuals. It allowed Van Wagenen to conduct pioneering studies on questions of reproduction, fertility, and endocrinology among humans and primates based on research on a model animal, the rhesus macaque. She dedicated more than forty years to this work. It would, however, lead to a radical change in the lives of hundreds of macaques, since her research included periodical anatomical studies, operations, and taking samples. In addition, the young were taken from their mothers at a very early age, as was the custom at the time.

Van Wagenen gathered data in this way on more than twelve hundred primates, assisted at more than six hundred births, and studied her charges over the course of fifteen generations. She also showed that the reproductive life (onset of menstruation, cycle of reproduction, fertile period, and menopause) of female rhesus macaques closely resembled that of female humans, even concerning the effects of hormones on them. Finally, it was thanks to Van Wagenen that American researchers in the 1950s became aware of the primatological studies being done by the Japanese. Traveling often, especially to observe macaque colonies in different countries, she had found one of Junichiro Itani's books, *Japanese Monkeys in Takasakiyama* (1954), in a Japanese bookstore and sent it to Stuart Altmann, an American biologist.

Apes and Anthropologists

By 1965, more than fifty researchers (primarily anthropologists and zoologists) were observing primates in their natural environment. The discipline of primatology saw a remarkable period of development occurring at the same time as the second wave of European and American women's movements. Specializing in physical anthropology, many women figured prominently in the tight circle of pioneers in field primatology.

Frances D. Burton, for one, explored the biological bases of the behavior of different species of primates in Costa Rica, Honduras, Barbados, China, Malaysia, and Morocco. She was the first to show that nonhuman female primates (such as macaques) could have orgasms.[17] A student of Sherwood Washburn, Phyllis Carol Jay (later Phyllis C. Dolhinow) examined infant behavior and mother-infant relations among Hanuman langurs.[18] She was among the woman primatologists who tried to redefine the role of females within primate societies.[19] Until this time, Western researchers had concentrated on male behavior, considering females passive and socially insignificant. Various prejudices linked to differences between the sexes influenced their observations and oriented their discourse on primates. Primate communities were thus often the objects of descriptions that were not only anthropocentric but also gendered. Jay defended the idea (audacious at the time) that females were at the heart of the social organization of the troops of langurs.

Another student of Washburn, Suzanne Ripley, conducted research on gray langurs in their natural environment and created a research site at Polonnaruwa (Sri Lanka) in the 1960s. She published a fundamental article in the history of primatology in 1967, in which she considered evolution, ecology, and behavior in a perspective that made indispensable the field studies of populations.[20]

Jane Beckman Lancaster (likewise a student of Washburn) was interested in the evolution of both human and nonhuman primates from an interdisciplinary perspective looking across different species.[21] She also proposed a new vision of female primates: based on long years spent in the field, Lancaster affirmed that female primates were not the submissive and passive beings portrayed by other researchers but were competitive and socially assertive. She showed that females did everything that the males did, created themselves opportunities for reproduction, played a fundamental social and cultural role, and developed tactics to challenge the strength of males.

A doctor in anthropology and professor at Yale University from 1972 to 1991, then the second woman vice chancellor at Cambridge University from 2003 to 2010, Alison Richard conducted field research in Central America, West Africa, the Himalayas, and most notably in Madagascar, where she worked on the population dynamics and social behavior of sifakas from an ecological perspective.[22]

The World from the Women's Point of View

Apart from these researchers who chose physical anthropology, other women, trained in biology, also left for the field. At Cambridge University, Thelma Rowell was enlisted by Hinde to study the relations between mothers and infants in the rhesus macaque colony at Madingley Laboratory. In 1962, she went to Uganda with her husband, the neuroethologist C. R. Fraser Rowell. Hired by Makere University in Kampala, she worked with captive baboons. She alternated her work at the university with research in the field. Alone in the field for two weeks of every month for over five years, Rowell studied the olive baboons of Ishasha Forest in Queen Elizabeth Park. She gradually habituated them to her presence, sleeping near them in a tent.

At the start of her research, she expected to observe very hierarchical societies dominated by males as well as numerous acts of aggression. She found, however, a much more peaceful social organization than she had thought possible. Passing constantly from one troop to another, and thus not constituting the stable core of the group, the males had to integrate into already-formed communities. They initially attempted to get themselves accepted by females on the periphery of the group, helping them take care of their infants and showing other signs of friendship.[23] The females thus played a central social role. In addition, the oldest females guided their communities to feeding grounds.

Contrary to accepted beliefs, the reproductive success of males was not necessarily linked to their hierarchical rank. Although this is true for some species, it is not proven for others. Rowell thus refuted the social insignificance of females, until then reduced to their sexual and maternal functions. At a time when the work of men implicitly constituted the norm and incarnated the scientific spirit, she denounced the male bias inherited from past studies.

Throughout the 1960s and 1970s, a number of women primatologists lent support to Rowell's position, including Jeanne Glaser Altmann, Shirley Strum, and Barbara Smuts. Jeanne Altmann formed a creative couple with Stuart Altmann, who played a determinant role in her life from the beginning of their research. In 1963 they decided to set up camp near Amboseli National Park in Kenya, at the foot of Mount Kilimanjaro. They stayed there for one year and wrote about their observations in a book, published in 1970, *Baboon Ecology: African Field Research.* In 1971, they founded the Amboseli Baboon Research Project. Jeanne Altmann showed that female baboons had "dual careers," juggling different activities while also taking care of their infants.[24]

At the beginning of the 1970s, Altmann played a large role in the domain of in situ studies. While this kind of research was experiencing a major development, scientists shared neither common methods nor rules. Aware of this problem, Altmann proposed various data-recording techniques, sampling methods, and methods of analysis. Profiting from her training in mathematics, she introduced statistics and mathematical models to the handling of data. The field of primatology was transformed. A feminist, Altmann also questioned the relation of women to the sciences. The American editor of a prestigious international publication in ethology, *Animal Behavior*, from 1978 to 1983, she contributed to the reconfiguration of the discourse on primates, questioning many overused terms, formulations, and categories.

The Baboons of Shirley Strum

Toward the end of 1972, Shirley Carol Strum began observations at Kekopey Ranch, northwest of Nairobi, near the village of Gilgil in the Great Rift Valley.[25] On her arrival, she was asked to study the olive baboons at a distance, from inside a bus, so as not to disturb the group and to protect herself from the potentially dangerous males. Believing that these studies were skewed because of the distance imposed between researchers and baboons, Strum left the bus and gradually habituated the primates to her presence. After an initial period spent learning to recognize each member of the group, she later consecrated a precise amount of time observing different individuals, male and female, adults and infants, dominant and nondominant. She was thus able to note the behavior of a representative sample of baboons.

As had other primatologists, she noticed things that were different from what she had learned at university. Far from being aggressive, the males spent most of their time watching what was happening around them, always on the lookout for changes that might occur in the group and upset its balance. The Pumphouse troop was not exclusively ruled by the brutal behavior of a handful of vindictive males. Solidarity between individuals was crucial for the survival of the group, contributing to the avoidance of attacks. Grooming, the reaffirmation of ties, cooperation, and goodwill were indispensable in maintaining social cohesion, itself linked to the construction of strong ties of confidence. The most aggressive baboons seemed to obtain less food and copulate less often than those that showed themselves

Shirley Strum in the world of baboons, Mukogodo, Laikipia (Kenya), Uaso Ngiro Baboon Project.
(Photograph courtesy of David Western)

to be peaceful and experienced, and capable of creating close ties of friendship with the females.[26]

Strum described sophisticated social abilities, irreducible to simplistic binary categories such as "dominance" versus "submission." It was thus necessary to think of interactions within the baboon troop in terms of a diversity of roles and keep the importance of aggressive behavior in perspective. Speaking of "complementary equality," Strum showed that if the idea of dominance has a place, it is practiced within the very hierarchical group of female baboons.

Lancaster, Altmann, Rowell, and Strum thus demonstrated that far from being perpetual victims and socially insignificant beings, female primates were fine strategists and highly competitive partners. Possessing a sharpened sense of hierarchy, they excelled in the management of complex social networks.

Women, Primates, and Sociobiology

In 1976, another woman became interested in baboons, the anthropologist and psychologist Barbara Smuts (née Boardman). A student of Robert

Hamburg (who collaborated with Jane Goodall) at Stanford University, she began her research on chimpanzees. She was among the sixty students (60 percent of whom were women) who worked at Gombe between 1966 and 1975. She was unfortunately kidnapped, along with three other students, by guerrillas active in the region. Once freed, she returned to the United States and pursued a PhD at Harvard University, where she was influenced by the ideas of Irven DeVore and Robert Trivers. She then left to study olive baboons in Tanzania and Kenya for two years, at the same sites as Shirley Strum.[27]

Smuts conducted one of the first sociobiological studies of primates, interpreting her observations in terms of fitness, costs, benefits, time and energy budgets, and strategies of optimization. The goal, for females as well as males, was to augment their reproductive success as well as their chance for survival. She also showed that the males who had the greatest chance of mating with females were those that established forms of "friendship" with them. Smuts noted that each baboon experienced this type of "amical partnership" in his or her own fashion, expressing a sense of "personalization" of primates supported by many woman primatologists.

Smuts's objective was to clarify the behavior of ancient hominids using the baboons as primatological models of human origins. Because of her field experience, she was in a position to question certain traditional evolutionary scenarios, showing, for example, that intimate ties between females and males could be established in the long term without exclusive sex, food sharing, sexual division of labor, or a nuclear family organization. The benefits of these associations between females and males were more social than economic. Smuts would go on to coedit the influential book *Primate Societies.*

The early work of these women primatologists was harshly criticized. Specialists in baboons quickly attacked their abilities and research methods. Established researchers, all-powerful in peer-reviewed literature, tried to lock out or slow down the spread of these new ideas. Nonetheless, Sarah Blaffer Hrdy explained in 1981 that the sexually passive, noncompetitive, and all-nurturing females of prevailing representations could not have evolved within primate societies. Gradually, it came to be admitted that there was a certain flexibility at work within primate troops. In 2006,

Shirley Strum and Bruno Latour described baboon societies in perpetual redefinition, in the sense that individuals, testing the availability and solidity of alliances, were constantly searching for foreseeable indications of future interactions. Far from being irremediably subject to a model centered around domination and aggressive behavior, the baboons themselves (female and male) subtly negotiated the core and contours of a fluid social organization.

Whatever Females Want, Females Get

In this necessarily limited survey of the first women primatologists, one cannot ignore another key researcher, Alison Bishop Jolly. Although not a specialist in the great apes, as were many of the women described in this chapter, Jolly nonetheless greatly influenced the field of research linked to the great apes by introducing a new paradigm to primatology. Fascinated by natural history since her childhood, Jolly received her PhD from Yale University in 1962 and began her field research the following year at Berenty Reserve in the south of Madagascar.[28] She would study lemurs for more than thirty years. By 1966 she showed that among ring-tailed lemurs, the females were "dominant," which went against the principal theories of the time about primate societies. The females controlled the terrain, were social leaders, and had priority over the resources: as Jolly wrote, whatever they wanted, they got.

Jolly was also interested in questions of sexuality, social intelligence, and cooperation from an evolutionary perspective.[29] She played an important role in the development of field research methods and was a pioneering environmental activist and key figure in lemur conservation. A new species, the Jolly's mouse lemur (*Microcebus jollyae*), was named to honor her as a prominent and inspiring figure in primatology.

Primatology and Underrepresented Researchers

In *Primate Visions,* Haraway notes that the majority of women primatologists are American and white. She says it is regrettable that so few minorities are represented and mentions one of the few black American primatologists, Clara B. Jones, little known despite her numerous publications. Jones's thesis, which she defended at Cornell University in 1978, focuses on the question of reproduction among mantled howler monkeys in Central America. Jones has worked in laboratories and in zoos as well as the field.[30] Today, fortunately, both African and Asian primatologists are better represented and have become more and more numerous.

In this gallery of portraits, we must not forget European researchers, of whom I will give but one example, the French woman Annie Gautier-Hion, specialist in the ecology of primates. Forming a "collaborative couple" with Jean-Pierre Gautier (in 1965), she traveled to Gabon in 1964 to observe gorillas.[31] She later studied a number of primate species, including a number of Old World monkeys such as Talapoins, which she made the object of her thesis in 1972.[32] Her work showed that the socioecological context was fundamental. Studying primates in conjunction with the tropical forest ecosystems they were part of, she showed that fruiting seasonality influenced the different methods of searching for food as a function of species, age, and sex. She defended the concept of "key resources" in the sense that the survival of primates depends crucially upon specific resources, especially during the dry season. In 1995, Annie Gautier continued her research of lowland gorillas in northern Congo-Brazzaville, where she worked for over ten years. She died in July 2011.

Primate Colonies

Within this panorama we must also make place for the women who founded the first colonies of great apes at the beginning of the twentieth century, three of whom have already been mentioned: Rosalía Abreu, Gertrude Lintz, and Maria Hoyt. These strong personalities were moved not only by the desire to own primates but also to educate and study them, even to teach them to speak. Wanting to advance science and share her knowledge of primates, Rosalía Abreu established an important correspondence with a number of American and European scientists, inviting some of them to her home in Cuba. She was also the first to assist at the birth of an ape in captivity, in 1915, and she studied in detail the development of a young ape.

Belle Jenning Benchley was also fascinated with great apes. In 1927, she became one of the few people who, like William Hornaday, Karl Marbe, Oskar Heinroth, or Anton Frederik Johan Portielje, conducted studies of captive primates. Having daily contact with the apes at the San Diego Zoo (gibbons, orangutans, chimpanzees, and gorillas), she observed them, organized experiments, and gathered knowledge. She was assisted by another woman, much less known, the scientist Joan Morton Kelly, a zoologist and former student of Carpenter. Benchley also collaborated with Ada and Robert Yerkes as well as Bingham and Carpenter. She maintained a correspondence with Rosalía Abreu and went to visit her colony. Benchley had a clear preference for gorillas. She said, however, that her favorite ape was the little chimpanzee Georgie, born at the zoo in 1938 and raised by his mother Katie. Not being a scientist, Benchley never became the official director of the zoo, even though she really was in charge. A great connoisseur of different exotic species, she gave a number of public lectures and was among the few women to publish books on primates before World War II.

Sanctuaries

Many women also founded sanctuaries for great apes, either in the primates' homeland or elsewhere. In Africa or Asia, they took in primates seized by authorities in the fight against illegal wildlife trafficking, those suffering from disastrous conditions of captivity in local zoos, young apes that had experienced bad treatment as pets, and orphans whose mothers had been killed by traffickers. Moved by the primates' plight and the need to find a refuge for them, these women initially started small institutions that could take in only a few individuals. As demand grew, however, they had to adapt. They became businesswomen, obtaining financial support and founding recognized international centers, where everything was set up for their protégés to lead peaceful lives. They took on every role: manager, accountant, human resource director, medical caregiver, architect, teacher, public relations worker.

Some of these women collaborated with scientific teams that came to conduct respectful and noninvasive research at their sanctuaries. Others tried to reintroduce their charges to their natural environment, with the help of local populations putting in place "integrated conservation" projects that combined conservation with social and economic development. Despite the complexity of this type of operation, the directors of these sanctuaries were for a long time looked down on and treated as sentimental empty nesters (the "big brown eyes" hypothesis) motivated by a mothering instinct, the primates substitutes for babies.

Rehabilitating Great Apes

Although the first rehabilitation center of this kind was created by a man, Eddie Brewer, in 1974, the majority of the founders or directors of sanctuaries are women, among them some primatologists (notably Jane Goodall).[33] The Chimpanzee Rehabilitation Trust founded by Eddie Brewer was taken up by his daughter, Stella Brewer-Marsden. A volunteer at the Nsele Zoo near Kinshasa for a number of years who had taken care of many orphans, Graziella Cotman founded the Tchimpounga Sanctuary (currently managed by the Jane Goodall Institute) in the Republic of Congo in 1981. Aliette Jamart, the director of the Association Habitat Écologique et Liberté des Primates (HELP Congo, founded in 1990), started a rehabilitation center in the same country in 1989. Orphaned chimpanzees were cared for and fed on three islands in the Conkuati lagoon. Jamart's ultimate goal, however, was to release the primates back into nature. She therefore put in place a reintroduction program in 1996 at a site in the Triangle Forest within the Conkuati-Douli National Park.

A volunteer at the Kinshasa Zoo, Claudine André met the young bonobo Mikeno in 1993, when she saved him from certain death. This experience changed her life. In 1994 she founded the Bonobo Sanctuary at Kinshasa. In 2002, she created a vast center at the *petites chutes* (small waterfalls) on the Lukaya River, about a half an hour by car from Kinshasa. Lola ya Bonobo is the only sanctuary in the world dedicated to this species, for which André has become the worldwide ambassador. She also initiated a rehabilitation program which, since 2009, has released a number of bonobos into a forest situated near the village of Bansankuru, whose inhabitants have become the "guardians of the bonobos," as part of an integrated conservation and development project.

Of the twenty-some sanctuaries in Africa that care for great apes, more than half were founded or are directed by women, and nearly a quarter by couples. The proportions are similar for the ten or so refuges that exist in Asia, in Borneo and Sumatra. In the United States, sanctuaries offer a peaceful retirement for primates that have undergone the worst kinds of trials in biomedical laboratories, participated in various research programs (such as at NASA), been kept as pets, or were used in the entertainment industry as performers in films, advertisements, or shows. Out of a dozen such sanctuaries, more than half were founded by women or are directed by them. The four largest centers (in numbers of primates) are run by women: Save the Chimps (founded by an exceptional woman, Carole Noon), Chimp Haven, Center for Great Apes (directed by Patti Ragan), and Project Chimps (which expects to take in over 250 chimpanzees).

The Beauties and the Beasts

Formerly portrayed as victims of male gorillas or orangutans, thrust into impossible love stories between Beauty and the Beast, women became emblematic protectors of great apes during the 1960s, a time of the loss of virile heroism during the period following the independence of Indonesia and the decolonization of Africa. Reports of their adventures were published in popular magazines. This strong presence in mass media suggested to the public that primate studies were typically done by women in the field. Extremely popular, the chimpanzees of Gombe and the mountain gorillas in Rwanda became consumer objects as well as exotic partners in the return to nature promoted by the hippie movement.

Apart from these images and slogans, women played a determinant role in launching long-term field studies and the construction of knowledge about primates. They initiated a change in methods, practices, and ways of looking at primates. Remaining for extended periods of time in the field, they observed intimately and were present throughout the lives of the primates and their descendants over many years. Furthermore, habituating primates to their presence permitted women to observe them up close, allowing them to record detailed observations and measure the sophistication of their societies. Female scientists also adopted a larger perspective taking into account ecological dimensions, conditions of existence, and exchanges between troops that lived in the same region.

In addition to this particular viewpoint, open to complexity, women were more sensitive to the effects of the male bias that had long affected observations. Being discriminated against themselves, they were able to notice how the role of female primates had been minimized or even ignored completely. They thus brought new attention to females in primate societies and questioned a number of stereotypes, especially those tied to the supposed "nature" of males and females or the sexual division of labor. They denounced the projection of patriarchal terminology and conceptions onto primate societies.

For Donna Haraway, the work of women primatologists has another common characteristic: their skepticism toward generalization. They believed that primates should not be thought of "in general": they should be seen as inscribed in a particular history, attached to a specific socioecological context, and actively related to a "real world" characterized by economic, cultural, and political issues. Alison Jolly stated that it was mostly due to the influence of women that a personalization of primates took place and an individualization of behavior became apparent. Added to this was women's attention to how natural environments changed and their recognition of the threats hanging over species, biotopes, and large forests. It would, however, be somewhat too simplistic to trace a strict line of demarcation between women and men, and unfair to attribute only to the former a role in renewing primatology and changing the way we look at apes.

Conclusion

Becoming-Human, Being-Ape

SHOWING TROUBLING SIMILARITIES TO humans while assumed to be of a "lower" order, ferocious and uncontrollable, driven by their primary urges, the great apes have throughout history aroused contradictory feelings between fascination and repulsion. They have long been seen as "in-between" beings, characterized by a constitutional ambivalence. Fears of a confusion with the animal and the possibility of their own regression toward bestiality have pushed humans to reject apes as belonging to nature, savage and uncultured. There have been, however, places and occasions when great apes have shown themselves to be incredibly close to us, beyond anything we may have imagined. What do these similarities between humans and apes tell us? Above all, they reveal in a striking manner the depth of our continuities. They have also certainly played a major role in the elaboration of ties between humans and apes, the former having undoubtedly been particularly sensitive to the efforts of the latter to imitate them and the resultant "mirror" effect.

Thus, the physical and behavioral similarities have captivated humans since antiquity, impelling them to bring primates into their homes and share their daily routines. These same similarities were later denounced as deceptions and imposture during the Christian Middle Ages. From antiquity onward, dissections and anatomical studies have extended this resemblance to internal structures. In the twentieth century, research in comparative psychology and later long-term field studies brought to light common psychic, social, and cultural traits. Beginning in the mid-twentieth century, the learning of human symbolic language by great apes attests to the presence of cognitive similarities between humans and apes, giving nuance to the primacy of the physical resemblance.

Furthermore, the capacity for apes integrated into families to appropriate human habits, savoir-faire, and ways of life has brought attention to one of their most remarkable characteristics: behavioral flexibility, defined by Clara Jones as a "toolbox of potential responses over time and space." This specific behavioral flexibility, accompanied by individual plasticity in the acquisition of a daily way of life, reinforces the learning capabilities of young apes. It allows for the emergence of innovative behaviors and the exercise of a certain degree of liberty.

Primates have not been content with simply integrating into human worlds; they have also incorporated ways of being in the world, traces of human ethos, experiencing impressive forms of what I call "becoming-human," whose depth surprises us since we do not expect it from the "world-poor" animals described by Heidegger. We are also destabilized by finding in the *Other* properties that belong to the *Same* to an unimaginable degree. Primates exhibit traits we had thought were exclusively human and arising from our culture: aesthetic sensibility, inventiveness, a capacity for symbolic thought (they "make believe," tell themselves stories, talk to their toys in sign language), the ability to wonder at a star-filled sky or admire a beautiful sunset. One of the most fascinating aspects of this continuum is self-consciousness, manifested in the most compelling manner, it seems to me, in the communities of humans and apes rather than in "mirror tests" (where self-representation and self-consciousness may be confused): such as when Viki the chimpanzee recognized her own photograph and classified it among photos of humans, not animals.

With Linnaeus, humans had feared for their own dissolution into the animal realm. With Darwin, they were terrorized by the idea of having a common ancestor with apes. In response to the dangerous proximity of great apes, humans have thus throughout time tried to accentuate the differences in order to deny the resemblances. Up until the twentieth century, the accent was placed on the intellectual and spiritual discontinuities. Those things seen as proper to humans also multiplied: the use of tools, politics, ethics, laughter, language, walking upright. This history has, however, shown that these traits can also be seen in our closest cousins to varying degrees. Recent research has shown cognitive characteristics in common: the capacity to coordinate different actions as means to an end, the functions of different areas of the brain, computational complexity, specialization of cerebral hemispheres, theory of mind, social intelligence, Broca's area, and the FOXP2 gene (associated with speech and language).[1]

We have seen throughout this short history that great apes are indicators par excellence of the limits and aporias of the rigid boundary raised between humans and animals. It is, furthermore, the apes themselves that deepen the similitudes and render this frontier obsolete, for example, when they designate themselves as human, walk upright on two feet, or carefully trace signs as if writing. Roger Fouts said of the female chimpanzee Washoe that to ask if she was ape or human when he held her simply held no sense for him. Maurice Temerlin said the same of Lucy. Janis Carter experienced a veritable "becoming-chimpanzee" during the period she lived among the rehabilitant apes on an island in Gambia, while the female chimpanzee Lucy continued to manifest a form of "becoming-human" she had picked up in the United States.

These apes and their human partners thus put into question the function of otherness that our closest phylogenetic cousins often take on. This otherness goes to the heart of representations tied to the relations between humans and other hominoids. The construction of the Other requires a dualist framework founded on the differentiation between the "same" and the "different." The great apes constitute this radical otherness. They allow humans to define themselves, to determine their properties, and to delimit their indescribable "humanity," removing it from the great abstract category of "animals."[2] Humans declare themselves to be radically separated from and superior to all other living beings. The rupture thus established validates a willful distancing from nature, a distancing lived as a means of resolving our existence both as "beings of nature" and "beings of culture."

Humans experience difficulties thinking of themselves both as an animal and as something "other" than animal, a position at odds with classical logic and the principle of noncontradiction at the center of rationalist thought. In a movement of polarization, they thus repress their animal side and try to distance themselves from all risk of contamination by that which, in their minds, is strongly linked to this category: the savage, the irrational, the primitive, and the bestial. "Finally there is in identity a double usage of difference. An affirmative use: the same persists in its own differentiating power. It is a creation. A negative use: the same defends itself from corruption by the other. It wants to preserve its purity. All identity is the dialectic game of a movement of creation, and a movement of purification."[3] The process of purification in which humans engage requires the extirpation of animal strata deeply embedded in the sediments of our being. Corporeal and material, traditionally opposed to the mind and soul (both immortal), this wild and savage part carries with it, irremediably, its inscription within temporality, and thus mortality, which makes the procedure all the more urgent.

The history of relations between humans and great apes is thus also one of strategies motivated by an obsession with purity, fear of chaos, and terror at being stained by the material. It is a history inscribed in Western dualist logic, which separates beings and their attributes into categories forced toward opposite and impenetrable poles, irreducible dichotomies: bestiality and humanity, nature and culture, wild and domesticated, primitive and civilized. Manifested in such a strong and irrefutable manner, the "becoming-human" of great apes threatens the identity thus constructed and orients the debate toward similitude rather than dissimilitude. We define ourselves as much by our proximities as by our differences. This is not to say that humans and great apes are interchangeable, nor to ignore the limits of the becoming-human of apes and the becoming-ape of humans. It therefore seems to me fruitful to think of our relations with great apes by adopting a position that recognizes the fragile and porous nature of the boundary separating them, taking into account both the continuities and the discontinuities without placing them in a hierarchy, seeing them as already inscribed within an essential common world (rather than within an artificially constructed boundary and a logic of difference). This proposition, which tends toward neither the exaltation of humans nor their devalorization, is nevertheless particularly delicate for those societies strongly marked by a dualist compulsion, and thus by a vision obsessed with the quest for superiority and inferiority, a need to reinforce a scale of beings and "races."[4]

Primatologists state that it is the apes themselves that lead them to their greatest discoveries and cause them to question their prejudices. As Donna Haraway says: "The animals are active participants in the constitution of what may count as scientific knowledge. From the point of view of the biologist's purposes, the animals resist, enable, disrupt, engage, constrain, and display. They act and signify, and like all action and signification, theirs yield no unique, univocal, unconstructed 'facts' waiting to be collected. The animals in behavioral biology are not transparent; they are dense."[5]

The principal actors of this brief history, these exceptional beings, are rapidly disappearing. All species of great apes are threatened, some of them critically. Many people think it is already too late. The very thought of the irreparable loss their extinction would cause should push us to do all that is possible to save these fellow travelers on this earth and to respect their territories. Paola Cavalieri and Peter Singer have proposed attributing fundamental rights and legal protection to great apes. They justify the extension of the human moral community to include great apes with scientific data: our behavioral, cognitive, and genetic continuity (more than 98 percent of genes in common) as well as the fact that each ape has a distinct personality.

Victoria, Washoe, Panzee, Lucy, Koko, Chantek, Nim, Kanzi, Congo, Wattana, and many others have shown their profound interest in our practices and have become expert in them. Their attachment to us, their fascinating behaviors, their artistic productions, their joyful and serious activities all stem from an essential community: they reveal an unexpected dimension, indescribable, that responds to neither experimental paradigm, logic of measure, nor mathematical modeling. The question here is not whether or not their behaviors are "human"—if the language they use is really language or their productions are really art. Their plasticity, their aesthetic sensibility, their spirit of invention, their creativity, and their Funktionslust are signs of something more fundamental: they reveal a shared sensibility, a vital impetus and force of desire, a common world, and a shared openness to the world.

Notes

Chapter One. The Uncanniness of Similitude

1. Plato, *Hippias Major,* 364–65.
2. Aristotle, *The History of Animals.*
3. Morris and Morris, *Men and Apes,* 15.
4. Cicero, *De Natura Deorum,* book 1, chapter 35.
5. The common Latin noun *simia,* monkey, can also mean imitator. According to Horst Waldemar Janson, the word *simian* is related to "similitude," undoubtedly on the basis of this sentence from Ennius. See *Apes and Ape Lore.*
6. Pliny the Elder, *Natural History,* book 8, chapter 80.
7. See Camper, *Organs of Speech.*
8. Bondt, *Historiae Naturalis,* 85. This early book of tropical medicine, originally in four volumes, was written in 1631 but not published until 1642, with a second edition in 1658.
9. Primates rarely survived more than a few months of captivity. At the time, nothing was known of their diet, how to take care of them, or how necessary it was to protect them from the cold.
10. See Tyson, *Orang-Outang.*
11. Bory de Saint-Vincent, quoted in Stoczkowski, "Portrait," 151.
12. See Cuvier, *The Animal Kingdom.*
13. Buffon, *Histoire naturelle,* 590. Note that "orang outan" was a generic name for the apes known at the time, chimpanzees and orangutans.
14. Buffon, "Nomenclature des singes."
15. Buffon, *Histoire naturelle,* 40–41.
16. Ibid., 580.
17. Vosmaer, *Description,* 15.
18. Ibid.
19. Cited in Schiebinger, *Nature's Body,* 207.
20. Sebastiani, "L'orang-outan," 18.
21. See Blanckaert, "'Produire l'être singe.'"

Chapter Two. Skeletons, Skins, and Skulls

1. As we shall see, comparative anatomy, developed by scientists such as Claude Galien, Pierre Belon, and Edward Tyson, would have an important place in

this process. Georges Cuvier was one of its advocates, as were Richard Owen and Thomas Henry Huxley.

2. See the excellent article by Barsanti, "L'orang-outan déclassé."
3. Buffon, "Des singes," 58.
4. Audebert, *Histoire naturelle.*
5. Barsanti, "L'orang-outan déclassé," 82.
6. See Matthews, *Voyage to the River Sierra-Leone.*
7. See Elliot, *Review of the Primates.*
8. Coolidge, "A Revision of the Genus Gorilla."
9. See Lamarck, *Zoological Philosophy.*
10. See the description in Grandpré, *Voyage à la côte occidentale d'Afrique,* 29. This description is repeated in Coupin, *Singes et singeries.*
11. See Blanckaert, "La question du singe."
12. Gratiolet and Alix, cited in ibid., 127.
13. See ibid., 130.
14. The phrase is proposed by Montgomery in *Primates in the Real World,* 12.
15. Often attributed, incorrectly, to the wife of the bishop of Worcester in 1860, this reaction to Darwin's theories seems in fact to derive from a sermon by the pastor Robert Forman Horton. See Horton, *Verbum Dei,* 132.
16. See Bowdich, *Mission from Cape Coast Castle.*
17. Milne-Edwards, "The Young Gorilla," 128.
18. See Gould, *Mismeasure of Man.*
19. Hartmann, *Anthropoid Apes,* 293.
20. Aron, *Essais d'épistémologie,* 193.
21. See Avrillion, *Mémoires.*
22. Donovan, *The Naturalist's Repository.*
23. Frédéric Cuvier and Geoffroy Saint-Hilaire, *Histoire naturelle des mammifères,* 257.
24. Quoted in Meunier, *Singes,* 139.
25. Meunier, *Singes domestiques,* 203.
26. Wallace, *The Malay Archipelago.*
27. See Haraway, *Primate Visions.*

Chapter Three. Apes as Guinea Pigs

1. See Hochadel, "Watching Exotic Animals."
2. Darwin, *Descent of Man,* 130.
3. This canon, still taught today, explains, "In no case may we interpret an action as the outcome of a higher psychical activity, if it can be interpreted as

the outcome of the exercise of one which stands lower in the psychological scale." See Morgan, *Introduction to Comparative Psychology,* 1st ed., 53. However, it is often forgotten that in the second edition of the work, Morgan clarified his thinking and modified the principle: "To this, however, it should be added, lest the range of the principle be misunderstood, that the canon by no means excludes the interpretation of a particular activity in terms of the higher processes if we already have independent evidence of the occurrence of these higher processes in the animal under observation." See Morgan, *Introduction to Comparative Psychology,* 2nd ed., 59.

4. See Garner, *The Speech of Monkeys.*
5. See Dewsbury, "Comparative Cognition" and *Comparative Psychology.*
6. Skinner, "A Case History," 230.
7. Haraway, *Primate Visions,* 233.
8. Others had discovered cures through similar methods: the winner of the 1930 Nobel Prize in Medicine, Karl Landsteiner, for his research on polio, or Maurice Wakeman on yellow fever.
9. Haraway, *Primate Visions,* 19.
10. If one adds to these numbers chimpanzees used in the entertainment industry or in the pet trade as well as those living in zoos, between 1,750 and 2,000 chimpanzees are living in North America. See http://www.chimpcare.org/.
11. See U.S. Fish and Wildlife Service, http://www.fws.gov/endangered/what-we-do/chimpanzee.html.
12. Dewsbury, *Comparative Psychology,* 64.
13. At the beginning, the members of the Royal Prussian Academy of Science envisaged a more ambitious project. They wanted to found a center where different species of apes could be tested and compared: gorillas, chimpanzees, orangutans, and gibbons. They were content, however, to begin with a single species. They decided, moreover, to concentrate on research into behavior and intelligence, leaving to one side, for a time, anatomical and physiological studies.
14. Some say there were nine chimpanzees, while others think that some chimpanzees died and were replaced.
15. The term *insight* is a translation of the German word *Einsicht,* the use of which marked a radical change in perspective. It passes in effect from (describable) behavior to mental processes (and hypothetical causes) related to this behavior. It is difficult to translate that which is observed as "intelligence." In using that word, one is immediately in danger of projecting human mental states onto animals; to see them as "pensive," "reflective," or "attentive" would be equivalent to committing the error of anthropomorphism.

There are competing schools of research and paradigms as regards this. The notion of "insight" in essence allows evocation, in a roundabout manner, of the mental "black box" without speaking the taboo words "mental processes" or "intentionality." Insight refers to the different forms of animal intelligence that manifest themselves without need for conditioning or intensive training, omnipresent in behaviorism. See Ruiz and Sánchez, "Wolfgang Köhler's *The Mentality of Apes.*"

16. Drawing from psychology, biology, and philosophy, gestalt theory or gestaltism (the theory of form) presents a holistic vision of an individual's perception of the world. According to this theory, the processes of perception and mental representation spontaneously treat phenomena as entire structures and not as a simple addition or juxtaposition of elements.
17. Köhler, "The Mentality of Orangs," 225.
18. See Köhler, *The Mentality of Apes.*
19. See Ley, *A Whisper of Espionage.*
20. Some speak of ordinary sadism. See Haraway, "Metaphors into Hardware," in *Primate Visions.*

Chapter Four. Great Apes in the Eyes of Scientists

Many parts of this chapter owe much to Jacques Vauclair, former director of research, Department of Developmental and Differential Psychology, Université de Provence, Aix-en-Provence; Bernard Thierry, director of research, Department of Ecology, Physiology, and Ethology, Université Louis Pasteur, Strasbourg; and Dalila Bovet, associate professor, Laboratory of Ethology, Cognition, Développement, Université Paris Ouest Nanterre La Défense.

1. See Montgomery, *Primates in the Real World.*
2. Beginning in 1913, Yerkes conducted research, mostly on pigs and crows, at this house, where he regularly welcomed other researchers and students. Before working with primates, Yerkes studied many other species: crabs, crayfish, jellyfish, fish, frogs, birds, turtles, mice, earthworms, cats, and others.
3. Yerkes, "Creating a Chimpanzee Community," 219.
4. Yerkes, "Provision for the Study of Monkeys and Apes," 234.
5. See Wynne, "Rosalía Abreu."
6. Yerkes and Yerkes, *The Great Apes,* 584.
7. Later, the center would become the Yerkes Laboratories of Primate Biology in honor of its founder when Yerkes retired in 1941. When the station became part of Emory University in 1956, the name was again changed, to the Yerkes National Primate Research Center. Still active today, the research center was

moved to Atlanta in 1965, then given a field station in Lawrenceville, Georgia, the following year.

8. See Haraway, "A Pilot Plant," in *Primate Visions.*
9. See the preface by Frans de Waal in Kohts, *Infant Chimpanzee.*
10. See Rossiianov, "Beyond Species."
11. On the question of hybridization, see the troubling novel by Vercors, *You Shall Know Them,* later made into a film, *Skullduggery* (1970).
12. See Todes, *Ivan Pavlov.*
13. See Windholz, "Pavlov vs. Köhler" and "Köhler's Insight Revisited."
14. See Radick, *The Simian Tongue,* 215.
15. Isidore Geoffroy Saint-Hilaire was the first to use the word *ethology* (defined as "the study of the behavior of animals in their natural environment"). William Morton Wheeler popularized the term in 1902 in *Nature.*
16. In the United States, a specialized journal was launched in 1911, the *Journal of Animal Behavior.* It was the first to be exclusively dedicated to the topic.
17. See Thinès and Zayan, "F. J. J. Buytendijk's Contribution."
18. The book, *Aus dem Leben der Bienen,* would not be published in English until 1953, as *The Dancing Bees.*
19. Cited by Renck, *L'éthologie,* 93.
20. Kummer, "Le comportement social des singes," 260. See also W. D. Hamilton, "Altruistic Behavior."
21. See Rees, *The Infanticide Controversy.*
22. See Wilson, *Sociobiology.* A popular version of sociobiology is Dawkins, *The Selfish Gene.*
23. Although some, like Skinner, did not exclude the idea of cognition but refused the concept of "mind" as explanation.
24. See Nagel, "What Is It Like to Be a Bat?"
25. See Chalmers, *The Conscious Mind,* 29.
26. See Griffin, *The Question of Animal Awareness.*
27. Lestel, *Les animaux sont-ils intelligents?* 37.
28. See Dennett, "Intentional Systems in Cognitive Ethology."
29. See Hunter, *The Delayed Reaction in Animals.*
30. See Tinklepaugh, "Multiple Delayed Reaction."
31. See Tolman, *Purposive Behavior* and "The Determiners of Behavior."
32. David Premack and Woodruff, "Does the Chimpanzee Have a Theory of Mind?" It is important not to have too monolithic an idea of the "theory of mind" but to take into account the different forms it can assume as well as the gradations in mental states.
33. Beck, "Chimpocentrism."

34. See Gallup, "Mirror-Image Stimulation" and "Chimpanzees."
35. Consciousness, so hard to define, manifests itself in varying degrees (up to the most complex) and manners: introspective, mature, conceptual, with (or without) self-awareness, and so on.
36. For field research, see Matsuzawa, Humle, and Yukimaru, *The Chimpanzees of Bossou;* and for laboratory research, see Matsuzawa, *Primate Origins.*
37. The question of counting among primates has been raised since the 1930s. See Ryo Kuroda, "On the Counting Ability of a Monkey"; and Gallis, "Les animaux savent-ils compter?"
38. http://matsuzawa.kyoto.
39. See Matsuzawa, "Chimpanzee Ai."
40. The Dutch school of ethology is notably represented by Nikolaas Tinbergen, Gerard Baerends (a student of Tinbergen, University of Groningen), Jaap Kruijt (one of the first to conduct detailed analysis on the development of social behaviors, University of Groningen), and Piet Wiepkema (among the pioneers in studies of animal well-being and author of a theory of emotions).
41. On the question of emotions in animals, see Moussaief Masson and McCarthy, *When Elephants Weep.*
42. See de Waal and Lanting, *Bonobo.*
43. See Vauclair, *Animal Cognition.*
44. See Derrida, *The Animal.*
45. Cheney, Seyfarth, and Smuts, *Social Relationships,* 1364.
46. See Byrne and Whiten, *Machiavellian Intelligence.*
47. See de Waal, preface to Kohts, *Infant Chimpanzee;* and de Waal, *The Ape and the Sushi Master.*

Chapter Five. Apes That Think They Are Human

1. See Cunningham, "A Gorilla's Life."
2. "The Civilized Gorilla."
3. See Raven, "Meshie" and "Further Adventures of Meshie."
4. See Raven's film, *Meshie, Child of a Chimpanzee.*
5. See Plowden, *Gargantua.*
6. See Hoyt, *Toto and I.*
7. See Noell and Himber, *The History of Noell's Ark.*
8. See Lenain, *Monkey Painting.*
9. See Radick, *The Simian Tongue;* and Lestel, *Paroles de singes.*
10. See Kellogg and Kellogg, *The Ape and the Child.*

11. See Keith Hayes and Hayes, "Imitation"; and Cathy Hayes, *The Ape in Our House.*
12. See Temerlin, *Lucy,* xxi.
13. See Fouts and Mills, *Next of Kin.*
14. See Ann James Premack, *Why Chimps Can Read.*
15. See Patterson and Linden, *The Education of Koko.*
16. See Terrace, *Nim.*
17. See Parker, Mitchell, and Miles, *Mentalities of Gorillas and Orangutans.*
18. See Savage-Rumbaugh and Lewin, *Kanzi.*
19. See Bender and Bender, "Swimming and Diving Behavior." See also "Cooper the Chimp Learns to Swim," https://www.youtube.com/watch?v=o5i1xhqoG2Y.
20. See Power, *The Egalitarians.*
21. Humans, particularly those working in the field, do the same in regard to apes.
22. See Hornaday, *Minds and Manners of Wild Animals.*
23. The dog Laika was the first animal sent into orbit to return alive. However, she died several hours after the flight, undoubtedly from the stress and the overheating she endured caused by the malfunction of the thermal regulating system.

Chapter Six. Conquering the Field

1. Martinez-Contreras, "L'émergence scientifique du gorille," 399.
2. See Hornaday, *Two Years in the Jungle.*
3. It seems, moreover, that at this time of widespread racism his birth may have unjustly prejudiced others against him, as his mother was from a racially mixed family on Réunion.
4. A mammalian taxonomist at the Museum für Naturkunde der Humbold-University zu Berlin, Paul Matschie also described the other subspecies of the eastern lowland gorilla (the largest of the five subspecies of gorilla), the Grauer's gorilla (*Gorilla beringei graueri*) in 1914, as well as a subspecies of the western lowland gorilla, the Cross River gorilla (*Gorilla gorilla dielhi*), in 1904.
5. On the capture of gorillas in the wild and providing them for zoos, see Pouillard, "Vie et mort."
6. See Montgomery, *Primates in the Real World,* 39–40.
7. See, for example, Osa Johnson's children's book *Snowball;* and Martin Johnson's book based on the film *Congorilla.*

8. See Marais, *The Soul of the Ape* and *My Friends the Baboons.*
9. In fact, it is common to keep silent about the contributions of the numerous collaborators participating in scientific studies, particularly those who made field studies possible. Western researchers are seen as heroes setting out alone to confront the jungle, but in reality they were accompanied by dozens of porters, cooks, and many other collaborators, taxidermists, and various technicians. Experts were also recruited from among the local populations. Naturalists could not have conducted their research without these guides, camp boys, trackers, rangers, and hunters, who were indispensable in finding, following, observing, and eventually capturing, the animals to which they dedicated a part of their lives. Being "nonprofessionals," these people, despite their immense knowledge of nature, were rarely mentioned.
10. See Derscheid, "Notes sur les gorilles."
11. Very active in the field of conservation, Derscheid was, like Akeley, a major player in the creation of Albert National Park.
12. See Pitman, "The Gorillas of the Kayonsa Region."
13. See Haraway, "Teddy Bear Patriarchy," in *Primate Visions.*
14. See Raven, "Gorilla."
15. See Nissen, *A Field Study of the Chimpanzee.*
16. See Bingham, *Gorillas in a Native Habitat.*
17. See Montgomery, "Place, Practice and Primatology."
18. Washburn, twenty-six years old at the time, took part in the Asiatic Primate Expedition as Carpenter's technical assistant and worked with Harold J. Coolidge.
19. Washburn, *The New Physical Anthropology.*
20. Washburn and Lancaster, "The Evolution of Hunting."
21. See Morbeck, Galloway, and Zihlman, *The Evolving Female.*
22. Concerning Japanese primatology, see, for example, Ohnuki-Tierney, "Representations of the Monkey"; Fedigan and Asquith, *The Monkeys of Arashiyama;* Matsuzawa and McGrew, "Kinji Imanishi"; and the work of Michael A. Huffman and Pamela J. Asquith.
23. This type of ritual (in the form of a burial or obituary in a newspaper) can also be found among other scientists, for example, Jane Goodall, Dian Fossey, and the researchers of the Amboseli Baboon Project.
24. See Asquith, "Primate Research Groups," 89.
25. Sugiyama, "Influence of Provisioning," 258.
26. A laboratory was created specifically for this group, the Koshima Field Laboratory (located at Kyushu, next to Koshima), under the aegis of the Primate Research Institute.

27. See Nishida, *The Chimpanzee of the Mahale Mountains.*
28. Two research centers were created at Kyoto University, the Primate Research Group, officially founded in 1951 by Kinji Imanishi and Denzaburo Miyadi and based at Kyoto, and the Primate Research Institute (or PRI), founded by Shunzo Kawamura and Masao Kawai in 1967 and based at Inuyama. The researchers of the two groups have close ties, but the PRI has developed an original perspective, somewhat divorced from the other. It rather quickly became the international center of primatological studies in Japan. See Asquith, "Primate Research Groups."

Chapter Seven. Socialities, Culture, and Traditions Among Primates

1. See Kummer, *Social Organization.*
2. Fossey, *Gorillas in the Mist,* 12.
3. Ibid., 11.
4. Following this research, in 1980 Robert M. Seyfarth, Dorothy L. Cheney, and Peter Marler would initiate the field of semantic communication in the animal world, with their studies (also done at Amboseli) of the different alarm calls of vervet monkeys, each call designating a different predator. The playback of previously recorded calls confirmed that each of them corresponded to a predator (such as leopard, eagle, or snake) and thus carried a specific meaning: the reactions of the primates were perfectly adapted to each call.
5. The idea that ancient hominids may have eaten insects as well has only recently been developed.
6. See Savage and Wyman, "Notice of the External Characters"; and Beatty, "A Note on the Behavior of the Chimpanzee."
7. See Goodall, *The Chimpanzees of Gombe.*
8. Fossey believed that the presence of tourists would carry the risk of miscarriages because of the stress, increase the possibility of disease transmission, and lessen the gorillas' distrust of humans (and thus of poachers).
9. See MacKinnon, *In Search of the Red Ape.*
10. See van Schaik et al., "Orangutan Cultures."
11. Van Schaik, *Among Orangutans.*
12. See Pradhan, van Noordwijk, and van Schaik, "A Model for the Evolution of Developmental Arrest."
13. See Hare, Wobber, and Wrangham, "The Self-Domestication Hypothesis."
14. See Furuichi and Thompson, *The Bonobos.*
15. See Beck, *Animal Tool Behavior.*
16. See Whiten et al., "Cultures in Chimpanzees."

17. See Kinani and Zimmerman, "Tool Use for Food Acquisition."
18. I have myself observed the use of sticks as weapons by bonobos at Planckendael Zoo (Belgium) in the early 2000s. Seeing a stork land on their island, a subadult male, Redy, ran toward it, followed by his two younger half brothers. He then picked up a long stick and threw it at the bird, without hitting it. He was imitated by Vifijo, a juvenile, who also threw a stick at the same target, though from farther away. The youngest, Zamba, only mimicked the gesture.
19. Tattersall, *Masters of the Planet,* 49–53.
20. Anne Russon and Biruté Galdikas have reported that some rehabilitated orangutans are able to light fires. Some keepers also tell of chimpanzees that light cigarettes without fear of the flame.
21. Perrot, *Mon histoire des femmes,* 160.
22. Knowledge of food constitutes an impressive body of learning: it seems that chimpanzees consume more than three hundred different kinds of food.
23. See Strum and Fedigan, *Primate Encounters.*

Chapter Eight. Women and Apes

1. See Haraway, *Primate Visions,* 293–95; see also Fedigan, "Is Primatology a Feminist Science?" 64.
2. See Jolly, *Lucy's Legacy,* 140.
3. Furthermore, while women are more numerous in primatology than in other branches of biology (for example, ornithology or mammalogy), there are proportionally no more women primatologists than there are women psychologists, anthropologists, or behavioral scientists. They are thus not overrepresented in professions tied to primates, although they are in many other domains, especially those professions associated with families, social services, child care, or care giving. They are also more numerous than men in the field of animal protection.
4. Mary French-Sheldon was the first white woman to undertake a solo expedition to Mombasa and Mount Kilimanjaro, at the beginning of the 1890s. She brought back a number of specimens as well as a wealth of knowledge about the fauna and local populations.
5. Traveling twice to Africa with her parents (in 1921 and 1924), Alice would go on to lead a life full of excitement following her early time in Africa. She was an artist, major in the U.S. Army Air Force, farmer, CIA employee, science fiction author, and novelist, publishing under the name James Tiptree Jr. She obtained a PhD in experimental psychology in 1967, when she was over fifty

years old. She presented herself as a man, driven to break free from a world dominated by men.

6. See, for example, Bradley and Bradley, "In Africa with Akeley."
7. Mary Bradley, *On the Gorilla Trail.*
8. She published J. T.'s biography (the first of a nonhuman primate) in 1928: *J. T., Jr.*
9. See Lykknes, Optiz, and Van Tiggelen, *For Better or for Worse?;* Pycior, Slack, and Abir-Am, *Creative Couples;* and Montgomery, *Primates in the Real World,* 135n8.
10. In a gender-structured society, women involved in research often have to arrange their time and resources within the framework of marriage. It should be noted, however, that the careers of primatologists are influenced not only by gender but also by social class and ethnic group.
11. Haraway, *Primate Visions,* 297.
12. On the question of the unequal distribution of resources, particularly for women in science, see Merton, "The Matthew Effect"; and Rossiter, "The Matthew Matilda Effect."
13. See Linden, "The Gambia."
14. Ibid., 37.
15. See Royer, "Facultés mentales."
16. Royer, "La domestication des singes."
17. Her publications include *Social Processes and Mental Abilities in Non-human Primates.*
18. A number of women primatologists (notably Zihlman, Lancaster, Jay, Ripley, and Strum) studied in the Department of Anthropology at the University of California at Berkeley under Washburn. The Subdepartment of Animal Behaviour at Cambridge, supervised by Robert Hinde and boasting a colony of rhesus macaques (Madingley Laboratory), was another important pole of primatology teaching. Goodall, Fossey, Cheney, and Rowell (who later went to Berkeley) all studied there. Utrecht University in the Netherlands was also a wellspring of primatologists in the department of the Van Hoof brothers, where the research was influenced by the work of Nikolaas Tinbergen. The 1970s saw the addition of programs in physical anthropology at Stanford University and Harvard University, where Irven DeVore taught his students a sociobiological perspective. See Haraway, *Primate Visions,* 300–302.
19. See Jay, "The Female Primate."
20. See Ripley, "The Leaping of Langurs."
21. See Lancaster, *Primate Behavior.*
22. See Richard, *Behavioral Variation.*

23. See Rowell, *Social Behavior of Monkeys.*
24. See Jeanne Altmann, *Baboon Mothers and Infants.*
25. A doctor of anthropology currently at the University of California at San Diego, Strum continues to work in the field. She has observed savanna baboons for nearly forty-five years as part of the Uaso Ngiro Baboon Project in Kenya's Laikipia Plateau.
26. See Strum, *Almost Human.*
27. See Smuts, *Sex and Friendship.*
28. See Jolly, *Lemur Behavior.*
29. See Jolly, *Lucy's Legacy.*
30. See Clara Jones, *Behavioral Flexibility.*
31. See Gautier, "Hommage à Annie Gautier-Hion."
32. See Gautier-Hion, Colyn, and Gautier, *Histoire naturelle des primates.*
33. See the site of the Pan African Sanctuary Alliance (PASA): https://www.pasaprimates.org.

Conclusion

1. As regards the first of these characteristics, at the Furuvik Zoo in Sweden, the male chimpanzee Santino (today thirty-three years old) collected stones in the water surrounding his island before the opening of the zoo. He then piled them only on the side where visitors passed, and only on days when the zoo was open. He also shaped stones into discs, which he then tried to throw at the visitors (happily, they were too far away for him to hit). He was thus perfectly capable of anticipating a situation, projecting his action into the future, formulating a strategy, and planning a series of acts with the goal of putting in place a certain line of attack.
2. See Derrida, *The Animal,* 2008.
3. Badiou, *Sarkozy,* 89.
4. See Fontenay, *Without Offending Humans.*
5. Haraway, *Primate Visions,* 310–11.

Bibliography

Akeley, Carl E. *In Brightest Africa.* Garden City, NY: Doubleday, Page, 1923.

Akeley, Carl E., and Mary L. Jobe Akeley. *Adventures in the African Jungle.* New York: Dodd, Mead, 1930.

Akeley, Delia. "Notes on African Monkeys," *Natural History* 18, no. 8 (1918).

———. *J. T., Jr.: The Biography of an African Monkey.* New York: Macmillan, 1928.

———. *Jungle Portraits.* New York: Macmillan, 1930.

Akeley, Mary L. Jobe. *Carl Akeley's Africa: The Account of the Akeley-Eastman-Pomeroy African Hall Expedition of the American Museum of Natural History.* New York: Dodd, Mead, 1929.

———. *Congo Eden.* New York: Dodd, Mead, 1950.

Altmann, Jeanne. *Baboon Mothers and Infants.* Cambridge, MA: Harvard University Press, 1980.

———. "Observational Study of Behavior Sampling Methods." *Behavior* 49 (1974): 227–65.

Altmann, Stuart A., and Jeanne Altmann. *Baboon Ecology: African Field Research.* Basel: Karger, 1970.

Alvey, Mark. "The Cinema as Taxidermy: Carl Akeley and the Preservative Obsession." *Framework: The Journal of Cinema and Media* 48, no. 1 (2007).

Ardrey, Robert. *African Genesis: A Personal Investigation into the Animal Origins and Nature of Man.* New York: Atheneum, 1961.

Aristotle. *The History of Animals.* Cambridge, MA: Harvard University Press, 1991.

Aron, Jean-Paul. *Essais d'épistémologie biologique.* Paris: Christian Bourgois, 1969.

Asquith, Pamela J. "Primate Research Groups." In *The Monkeys of Arashiyama: Thirty-Five Years of Research in Japan and the West,* edited by Linda M. Fedigan and Pamela J. Asquith. New York: State University of New York Press, 1991.

Audebert, Jean-Baptiste. *Histoire naturelle des singes et des makis.* Paris: Desray, 1799–1800.

Avrillion, Marie. *Mémoires de Mademoiselle Avrillion, première femme de chambre de l'impératrice: Sur la vie privée de Joséphine, sa famille et sa cour.* Paris: Mercure de France, 1969.

Badiou, Alain. *De quoi Sarkozy est-il le nom?* Paris: Nouvelles Éditions Lignes, 2007.

Baratay, Eric. *Biographies animales: Des vies retrouvées.* Paris: Éditions du Seuil, 2017.

Barsanti, Giulio. "L'orang-outan déclassé (Pongo Wurmbii Tied.): Histoire du premier singe à hauteur d'homme (1780–1801) et ébauche d'une théorie de la circularité des sources." In "Histoire de l'anthropologie: Hommes, idées, moments," edited by C. Blanckaert, A. Ducros, and J.-J. Hublin. Special issue, *Bulletins et mémoires de la Société d'Anthropologie de Paris,* n.s., 1, nos. 3–4 (1989): 67–104.

Battell, Andrew. "The Strange Adventures of Andrew Battell, of Leigh in Essex, Sent by the Portugails Prisoner to Angola, Who Lived There, and in the Adjoining Regions Near Eighteen Years" [1613]. *In Purchas His Pilgrimes,* 2nd ser., vol. 6. London: Hakluyt Society, 1901.

Beatty, Harry A. "A Note on the Behavior of the Chimpanzee." *Journal of Mammalogy* 32, no. 118 (1951).

Beck, Benjamin B. *Animal Tool Behavior: The Use and Manufacture of Tools by Animals.* New York: Taylor & Francis, 1980.

———. "Chimpocentrism: Bias in Cognitive Ethology." *Journal of Human Evolution* 11, no. 1 (1982).

Beeckman, Daniel. *A Voyage to and from the Island of Borneo.* 1718. Reprint, Folkestone, UK: Dawson's of Pall Mall, 1973.

Belloni Du Chaillu, Paul. *Explorations and Adventures in Equatorial Africa.* London: John Murray, 1861.

———. *Stories of the Gorilla Country Narrated for Young People.* New York: Harper & Brothers, 1868.

Benchley, Belle J. *My Friends, the Apes.* Boston: Little, Brown, 1942.

———. *My Life in a Man-Made Jungle.* Boston: Little, Brown, 1940.

Bender, Daniel E. *The Animal Game.* Cambridge, MA: Harvard University Press, 2016.

Bender, Renato, and Nicole Bender. "Brief Communication: Swimming and Diving Behavior in Apes (*Pan troglodytes* and *Pongo pygmaeus*): First Documented Report." *American Journal of Physical Anthropology* 152 (2013): 156–62.

Bingham, Harold C. *Gorillas in a Native Habitat: Report of the Joint Expedition of 1929–30 of Yale University and Carnegie Institution of Washington for Psychobiological Study of Mountain Gorillas (*Gorillas beringei*) in Parc National Albert, Belgian Congo, Africa.* Washington, DC: Carnegie Institution of Washington, 1932.

Blanckaert, Claude. "'Produire l'être singe': Langage du corps et harmonies spirituelles." *Annales historiques de la Révolution française,* July–September 2014, 377.

———. "La question du singe." In Corbey and Theunissen, *Ape, Man, Apeman.*

Boesch, Christophe. *Wild Cultures: A Comparison Between Chimpanzee and Human Cultures.* Cambridge: Cambridge University Press, 2012.

Boesch, Christophe, and Hedwige Boesch-Achermann. *The Chimpanzees of the Taï Forest: Behavioural Ecology and Evolution.* Oxford: Oxford University Press, 2000.

Bondt, Jacobus de. *Historiae Naturalis & Medicae Indiae Orientalis.* 6 vols. 2nd ed. In *De Indiae Utriusque Re Naturali et Medica,* by G. Piso. Amsterdam: Ludovicum et Danielem Elzevirios, 1658.

Boutan, Louis. *Pseudo-langage.* Bordeaux: A. Saugnac, 1913.

Bowdich, T. Edward. *Mission from Cape Coast Castle to Ashantee.* London: John Murray, 1819.

Bradley, Herbert, and Mary Hastings Bradley. "In Africa with Akeley." *Natural History* 27, no. 2 (1927): 161–73.

Bradley, Mary Hastings. *On the Gorilla Trail.* London: D. Appleton, 1922.

Breydenbach, Bernhard von. *Peregrinatio in Terram Sanctam.* Mainz: Erhard Reuwich, 1486.

Buffon, Georges-Louis Leclerc de. "Des singes." In *Histoire naturelle générale et particulière,* vol. 35. Paris: Dufart, 1800.

———. *Histoire naturelle: Quadrumanes.* Vols. 33–34 of *Œuvres complètes de Buffon avec les suites de Lacépède précédées d'une notice sur la vie et les ouvrages de Buffon par M. le Baron Cuvier.* Rev. ed. Paris: Lecointe, 1830.

———. "Nomenclature des singes." In *Histoire naturelle, générale et particulière, avec la description du cabinet du roi,* vol. 28. Paris: Imprimerie royale, 1766.

Burbridge, Ben. *Gorilla: Tracking and Capturing the Ape-Man of Africa.* New York: Century, 1928.

Burkhardt, Richard W. "The Founders of Ethology and the Problem of Human Aggression: A Study in Ethology's Ecologies." In *The Animal/Human Boundary: Historical Perspectives,* edited by A. N. H. Creager and J. W. Chester, 265–304. New York: University of Rochester Press, 2002.

———. *Patterns of Behavior: Konrad Lorenz, Niko Tinbergen, and the Founding of Ethology.* Chicago: University of Chicago Press, 2005.

Burton, Frances D., ed. *Social Processes and Mental Abilities in Non-human Primates: Evidences from Longitudinal Field Studies.* Lewiston, NY: Edwin Mellen, 1992.

Burton, Richard F. *Two Trips to Gorilla Land and the Cataracts of the Congo.* London: S. Low, Marston, Low, & Searle, 1876.

Byrne, Richard W., and Andrew Whiten. *Machiavellian Intelligence: Social Expertise and the Evolution of Intellect in Monkeys, Apes, and Humans.* Oxford: Oxford University Press, 1988.

Camper, Adrien G., ed. *Dissertation physique de Mr. Pierre Camper, sur les différences réelles que présentent les traits du visage chez les hommes des différents pays et de différents âges; Sur le beau qui caracterise les statues antiques et les pierres gravées.* Utrecht: B. Wild & J. Altheer, 1791.

Camper, Peter. "Account of the Organs of Speech of the Orang Outang." *Philosophical Transactions of the Royal Society of London* 69 (1779): 139–60.

———. *Histoire naturelle de l'orang-outang, et de quelques autres singes.* Harlingae: Van der Plaats, 1779.

Carpenter, Clarence R. *A Field Study of the Behavior and Social Relations of Howling Monkeys (*Alouatta palliata*). Comparative Psychology Monographs* 10, no. 2. Baltimore, MD: Johns Hopkins University Press, 1934.

———. *Naturalistic Behavior of Nonhuman Primates.* University Park: Pennsylvania State University Press, 1964.

Cavalieri, Paola, and Peter Singer. *The Great Ape Project: Equality Beyond Humanity.* New York: St. Martin's Griffin, 1994.

Chalmers, David J. *The Conscious Mind.* New York: Oxford University Press, 1996.

Cheney, Dorothy, and Robert M. Seyfarth. *How Monkeys See the World: Inside the Mind of Another Species.* Chicago: University of Chicago Press, 1990.

Cheney, Dorothy L., Robert M. Seyfarth, and Barbara B. Smuts. "Social Relationships and Social Cognition in Non-human Primates." *Science* 234 (1986).

Cicero, Marcus Tullius. *De Natura Deorum (On the Nature of Gods).* Translated by Francis Brooks. London: Methuen, 1896.

"The Civilized Gorilla." *Literary Digest,* December 10, 1921.

Claudius Aelianus. *Aelian's "On the Nature of Animals."* Translated by Gregory McNamee. San Antonio: Trinity University Press, 2011.

Coolidge, Harold J. *The Living Asiatic Apes: An Account of the Asiatic (Harvard) Primate Expedition.* Boston: Harvard Bulletin, 1938.

———. "*Pan paniscus:* Pigmy Chimpanzee from South of the Congo River." *American Journal of Physical Anthropology* 18, no. 1 (1933): 1–57.

———. "A Revision of the Genus *Gorilla.*" *Memoirs of the Museum of Comparative Zoology at Harvard College* 50 (1929): 295–381.

Corbey, Raymond, and Bert Theunissen, eds. *Ape, Man, Apeman: Changing Views Since 1600.* Leiden: Department of Prehistory, Leiden University, 1995.

Coupin, Henri. *Singes et singeries: Histoire anecdotique des singes.* Paris: Vuibert & Nony Editeurs, 1907.

Cunningham, Alyce. "A Gorilla's Life in Civilization." *Zoological Society Bulletin* 24, no. 5 (1921): 118–24.

Cuvier, Frédéric, and Étienne Geoffroy Saint-Hilaire. *Histoire naturelle des mammifères.* 2nd ed. Paris: Firmin Didot, 1819.

———. "Mémoire sur les orangs-outans." *Journal de physique, de chimie et d'histoire naturelle* 3, no. 46 (1798): 185–191.

Cuvier, Georges. *The Animal Kingdom.* London: Whittaker, Treacher, 1827.

Cuvier, Georges, and Étienne Geoffroy Saint-Hilaire. "Histoire naturelle des orangs-outans." *Magasin encyclopédique* 1, no. 3 (1795): 451–63.

Dahan, Amy, and Dominique Pestre. *Les Sciences pour la guerre, 1940–1960.* Paris: Éditions de l'EHESS, 2004.

Dart, Raymond Arthur. "The Predatory Transition from Ape to Man." *International Anthropological and Linguistic Review* 1 (1953): 201–17.

Darwin, Charles. *The Descent of Man, and Selection in Relation to Sex.* London: John Murray, 1871.

———. *The Expression of the Emotions in Man and Animals.* London: John Murray, 1872.

———. *The Origins of Species, by Means of Natural Selection.* London: John Murray, 1859.

Dawkins, Richard. *The Selfish Gene.* Oxford: Oxford University Press, 1976.

Deleuze, Gilles, and Félix Guattari. *Kafka: Toward a Minor Literature.* Minneapolis: University of Minnesota Press, 1986.

———. *A Thousand Plateaus: Capitalism and Schizophrenia.* London: Continuum, 1988.

Dennett, Daniel C. "Intentional Systems in Cognitive Ethology: The 'Panglossian Paradigm' Defended." *Behavioral and Brain Sciences* 6 (1983): 343–55.

Derrida, Jacques. *The Animal That Therefore I Am.* New York: Fordham University Press, 2008.

Derscheid, Jean-Marie. "Notes sur les gorilles des volcans du Kivu." *Annales de la Société Royale Zoologique de Belgique* 58 (1927): 149–59.

DeVore, Irven, ed. *Primate Behavior.* New York: Holt, Rinehart & Winston, 1965.

DeVore, Irven, and Sherwood L. Washburn. "Baboon Ecology and Human Evolution." In *African Ecology and Human Evolution,* edited by F. C. Howell and F. Bourliere. Chicago: Aldine, 1963.

de Waal, Frans. *The Age of Empathy.* New York: Crown, 2009.

———. *The Ape and the Sushi Master.* New York: Basic Books, 2001.

———. *Are We Smart Enough to Know How Smart Animals Are?* New York: Norton, 2016.

———. *Chimpanzee Politics: Power and Sex Among Apes.* London: Allen & Unwin, 1982.

———. *Good Natured: The Origins of Right and Wrong in Humans and Other Animals.* Cambridge, MA: Harvard University Press, 1996.

———. *Peacemaking Among Primates.* Cambridge, MA: Harvard University Press, 1989.

de Waal, Frans, and Frans Lanting. *Bonobo: The Forgotten Ape.* Berkeley: University of California Press, 1997.

Dewsbury, Donald A. "Comparative Cognition in the 1930s." *Psychonomic Bulletin & Review* 7, no. 2 (2000): 267–83.

———. *Comparative Psychology in the Twentieth Century.* Stroudsburg, PA: Hutchinson Ross, 1984.

Diamond, Jared. *The Rise and Fall of the Third Chimpanzee.* London: Radius, 1991.

Donisthorpe, Jill J. "A Pilot Study of the Mountain Gorilla (*Gorilla gorilla beringei*) in South West Uganda, February to September 1957." *South African Journal of Science* 54, no. 8 (1958): 195–217.

Donovan, Edward. *The Naturalist's Repository; or, Monthly Miscellany of Exotic Natural History.* Vol. 2. London: W. Simpkin & R. Marshall, 1824.

Elliot, Daniel G. *A Review of the Primates.* 3 vols. New York: American Museum of Natural History, 1913.

Fedigan, Linda M. "Is Primatology a Feminist Science?" In *Women in Human Evolution,* edited by Lori Hager. New York: Routledge, 1997.

———. *Primate Paradigms, Sex Roles and Social Bonds.* Chicago: University of Chicago Press, 1980.

———. "Science and the Successful Female: Why There Are So Many Women Primatologists." *American Anthropologist,* n.s., 96, no. 3 (1994): 529–40.

Fedigan, Linda M., and Pamela J. Asquith, eds. *The Monkeys of Arashiyama: Thirty-Five Years of Research in Japan and the West.* New York: State University of New York Press, 1991.

Fontenay, Elisabeth de. *Le silence des bêtes: La philosophie à l'épreuve de l'animalité.* Paris: Fayard, 1998.

———. *Without Offending Humans: A Critique of Animal Rights.* Minneapolis: University of Minnesota Press, 2012.

Fossey, Dian. "The Behaviour of the Mountain Gorilla." PhD diss., University of Cambridge, 1976.

———. *Gorillas in the Mist.* Boston: Houghton Mifflin, 1983.

———. "Making Friends with Mountain Gorillas." *National Geographic,* January 1970, 48–68.

Fouts, Roger, and Stephen Tukel Mills. *Next of Kin: What Chimpanzees Have Taught Me About Who We Are.* New York: William Morrow, 1997.

Frisch, Karl von. *The Dancing Bees: An Account of the Life and Senses of the Honey Bee.* New York: Harvest Books, 1953.

Fulton, John F. "The Chimpanzee in Experimental Medicine." *Transactions of the Kansas City Academy of Medicine, 1937, 1938, 1939, Section of Primate Physiology, Laboratory of Physiology, Yale School of Medicine, Collected Papers July 1, 1939–June 30, 1940.* Vol. 7, paper 239.

Furness, William H. "Observations on the Mentality of Chimpanzees and Orang-Utans." *Proceedings of the American Philosophical* Society 55, no. 3 (1916): 281–90.

Furuichi, Takeshi, and Jo Thompson, eds. *The Bonobos: Behavior, Ecology and Conservation.* New York: Springer, 2008.

Galdikas, Biruté M. F. "Orangutans, Indonesia's 'People of the Forest.'" *National Geographic,* October 1975, 444–73.

———. *Reflections of Eden: My Years with the Orangutans of Borneo.* New York: Little, Brown, 1995.

Gallis, P. "Les animaux savent-ils compter?" *Bulletin de la Société Royale des Sciences de Liège* 1 (1932): 82–84.

Gallup, Gordon G., Jr. "Chimpanzees: Self-Recognition." *Science* 167 (1970): 86–87.

———. "Mirror-Image Stimulation." *Psychological Bulletin* 70 (1968): 782–93.

Gardner, Howard. *The Mind's New Science: A History of the Cognitive Revolution.* New York: Basic Books, 1985.

Gardner, Allen R., Beatrix T. Gardner, and Thomas E. Van Cantfort, editors. *Teaching Sign Language to Chimpanzees.* Albany: State University of New York Press, 1989.

Garner, Richard Lynch. *Gorillas and Chimpanzees.* London: Osgood, McIlvaine, 1896.

———. *The Speech of Monkeys.* New York: Charles L. Webster, 1892.

Gautier, Jean-Pierre. "Hommage à Annie Gautier-Hion." *Revue de primatologie* 4 (2012).

Gautier-Hion, Annie, Marc Colyn, and Jean-Pierre Gautier. *Histoire naturelle des primates d'Afrique Centrale.* Libreville, Gabon: Ecofac Editions, 1999.

Geoffroy Saint-Hilaire, Étienne. "Considérations sur les singes les plus voisins de l'homme." *Comptes rendus hebdomadaires des séances de l'Académie des Sciences* 2 (1836): 92–95.

Gesner, Conrad. "Historiae Animalium Lib. I. de Quadrupedibus Viviparis." In *Historiae Animalium*. Tiguri: Apud Christ. Froschoverum, 1551–58.

Goodall, Jane. *The Chimpanzees of Gombe: Patterns of Behavior.* Cambridge, MA: Belknap Press, Harvard University Press, 1986.

———. "My Life Among Wild Chimpanzees." *National Geographic,* August 1963, 272–308.

Gordon, Nicholas. *Murders in the Mist: Who Killed Dian Fossey?* London: Hodder & Stoughton, 1994.

Gould, Stephen Jay. *The Mismeasure of Man.* New York: Norton, 1981.

Grandpré, Louis-Marie-Joseph Ohier, comte de. *Voyage à la côte occidentale d'Afrique.* 2 vols. Paris: Dentu, 1801.

Gregory, William K., and Henry C. Raven. *In Quest of Gorillas.* New Beford, MA: Darwin Press, 1937.

Griffin, Donald R. *The Question of Animal Awareness: Evolutionary Continuity of Mental Experience.* New York: Rockefeller University Press, 1976.

Guillaume, Paul, and Ignace Meyerson. *Recherches sur l'usage de l'instrument chez les singes.* Vol. 23. Paris: Vrin, 1987.

Hager, Lori, ed. *Women in Human Evolution.* New York: Routledge, 1997.

Hahn, Emily. *Eve and the Apes.* New York: Weidenfeld & Nicolson, 1988.

Hamilton, Gilbert W. "A Study of Sexual Tendencies in Monkeys and Baboons." *Journal of Animal Behavior* 4, no. 5 (1914): 295–318.

Hamilton, William D. "The Evolution of Altruistic Behavior." *American Naturalist* 97, no. 896 (1963): 354–56.

Haraway, Donna. *Primate Visions: Gender, Race, and Nature in the World of Modern Science.* New York: Routledge, 1989.

———. *Simians, Cyborgs, and Women: The Reinvention of Nature.* New York: Routledge, 1991.

Hare, Brian, Victoria Wobber, and Richard Wrangham. "The Self-Domestication Hypothesis: Evolution of Bonobo Psychology Is Due to Selection Against Aggression." *Animal Behaviour* 83, no. 3 (2012): 573–85.

Harlow, Harry F., and Margaret K. Harlow. "The Affectional Systems." *Behavior of Nonhuman Primates: Modern Research Trends* 2 (1965): 287–334.

Harrisson, Barbara. *Orang-Utan: The Story of Some Unusual Guests; Orang-Utans in a Private Household.* New York: Doubleday, 1963.

Hartmann, Robert. *Anthropoid Apes.* New York: Appleton, 1886.

Hayes, Cathy. *The Ape in Our House.* New York: Harper, 1951.

Hayes, Keith J., and Catherine Hayes. "Imitation in a Home-Raised Chimpanzee." *Journal of Comparative and Physiological Psychology* 45, no. 5 (1952): 450–59.

Herzfeld, Chris. "L'invention du bonobo." *Bulletin d'histoire et d'épistémologie des sciences de la vie* 14, no. 2 (2007).

———. *Wattana: An Orangutan in Paris.* Chicago: University of Chicago Press, 2016.

Herzfeld, Chris, and Dominique Lestel. "Knot Tying in Great Apes: Etho-ethnology of an Unusual Tool Behavior." *Social Science Information* 44, no. 4 (2005): 621–53.

Hess, Christine. *The Baby Chimp.* London: G. Bell and Sons, 1954.

Hinde, Robert Aubrey. *Animal Behaviour: A Synthesis of Ethology and Comparative Psychology.* New York: McGraw-Hill, 1966.

Hobhouse, Leonard Trelawney. *Mind in Evolution.* London: Macmillan, 1901.

Hochadel, Oliver. "Watching Exotic Animals Next Door: 'Scientific' Observations at the Zoo (ca. 1870–1910)." *Science in Context* 24, no. 2 (2011): 183–214.

Horn, Arthur D. "A Preliminary Report on the Ecology and Behavior of the Bonobo Chimpanzee (*Pan paniscus* Schwarz 1929) and a Reconsideration of the Evolution of the Chimpanzee." PhD diss., Yale University, 1976.

Hornaday, William T. *The Minds and Manners of Wild Animals: A Book of Personal Observations.* New York: Scribner's Sons, 1922.

———. *Two Years in the Jungle: The Experiences of a Hunter and Naturalist in India, Ceylon, the Malay Peninsula and Borneo.* New York: Scribner's Sons, 1885.

Horton, Robert Forman. *Verbum Dei: The Yale Lectures on Preaching, Lecture IV: The Bible and the Word of God.* New York: Macmillan, 1893.

Hoyt, Maria A. *Toto and I: A Gorilla in the Family.* Philadelphia: J. B. Lippincott, 1941.

Hrdy, Sarah Blaffer. *Mother Nature: A History of Mothers, Infants, and Natural Selection.* New York: Pantheon Books, 1999.

———. *The Woman That Never Evolved.* Cambridge, MA: Harvard University Press, 1981.

Hunter, Walter S. *The Delayed Reaction in Animals. Behavior Monographs* 2 (1912).

Huxley, Thomas Henry. *Evidence as to Man's Place in Nature.* London: Williams & Norgate, 1863.

Imanishi, Kinji. *A Japanese View of Nature: The World of Living Things.* London: Routledge Curzon, 2002.

Ingold, Tim. *The Perception of the Environment: Essays on Livelihood, Dwelling and Skill.* London: Routledge, 2000.

Itani, Junichiro. "Japanese Monkeys in Takasakiyama." In *Social Life of Animals in Japan,* edited by Kinji Imanishi. Tokyo: Kobunsha, 1954.

———. "Twenty Years with the Mount Takasaki Monkeys." In *Primates Utilization and Conservation,* edited by G. Bermant and D. Lindberg, 197–249. New York: Wiley-Interscience, 1975.

Jahme, Carole. *Beauty and the Beasts: Woman, Ape, and Evolution.* New York: Soho, 2001.

Janson, Horst Waldemar. *Apes and Ape Lore in the Middle Ages and the Renaissance.* London: Warburg Institute, University of London, 1952.

Jay, Phyllis. "The Female Primate." In *Man and Civilization: The Potential of Women,* edited by S. M. Farber and R. H. L. Wilson, 3–12. New York: McGraw-Hill, 1963.

Johnson, Martin. *Congorilla: Adventures with Pygmies and Gorillas in Africa.* New York: Blue Ribbon Books, 1931.

———. "Scenes from Akeley's Africa." *Natural History* 27 (1927): 172–73.

Johnson, Osa. *Snowball: Adventures of a Young Gorilla.* New York: Random House, 1942.

Jolly, Alison. *The Evolution of Primate Behavior.* New York: Macmillan, 1972.

———. *Lemur Behavior: A Madagascar Field Study.* Chicago: University of Chicago Press, 1964.

———. *Lucy's Legacy: Sex and Intelligence in Human Evolution.* Cambridge, MA: Harvard University Press, 1999.

Jones, Clara B. *Behavioral Flexibility in Primates: Causes and Consequences.* New York: Springer, 2005.

Jones, Clyde, and Jorge Sabater Pi. "Sticks Used by Chimpanzees in Rio Muni, West Africa." *Nature* 223, no. 5201 (1969): 100–101.

Kano, Takayoshi. *The Last Ape: Pygmy Chimpanzee Behavior and Ecology.* Stanford: Stanford University Press, 1992.

———. "A Pilot Study on the Ecology of Pygmy Chimpanzees, *Pan paniscus.*" In *The Great Apes,* edited by D. A. Hamburg and E. R. Mc Cown, 122–35. Menlo Park, CA: Benjamin/Cummings, 1979.

———. "The Use of Leafy Twigs for Rain Cover by the Pygmy Chimpanzees of Wamba." *Primates* 23, no. 3 (1982): 453–57.

Kawai, Masao. *Nihonzaru no seitai* [Life of Japanese Monkeys]. Tokyo: Kawadeshohoshinsha, 1969.

Kawai, Masao, and Hiroki Mizuhara. "An Ecological Study of the Wild Mountain Gorilla (*Gorilla gorilla beringei*): Report of the JMC 2nd Gorilla Expedition." *Primates* 2, no. 1 (1959): 1–42.

Kellogg, Winthrop N. "Communication and Language in the Home-Raised Chimpanzee." *Science,* n.s., 162, no. 3852 (1968): 423–27.

Kellogg, Winthrop N., and Luella A. Kellogg. *The Ape and the Child: A Study of*

Environmental Influence upon Early Behavior. New York: McGraw-Hill, 1933.

Kinani, Jean-Felix, and Dawn Zimmerman. "Tool Use for Food Acquisition in a Wild Mountain Gorilla (*Gorilla beringei beringei*)." *American Journal of Primatology* 77 (2015): 353–57.

Kingsley, Mary. *Travels in West Africa, Congo Francais, Corisco and Cameroons.* London: Macmillan, 1897.

———. *West African Studies.* London: Macmillan, 1899.

Kirk, Jay. *Kingdom Under Glass: A Tale of Obsession, Adventure, and One Man's Quest to Preserve the World's Great Animals.* New York: Henry Holt, 2010.

Köhler, Wolfgang. *The Mentality of Apes.* New York: Harcourt, Brace, 1925.

———. "The Mentality of Orangs." *International Journal of Comparative Psychology* 6, no. 4 (1993): 189–229.

Kohts, Nadezhda N. Ladygina. *Infant Chimpanzee and Human Child.* Edited by Frans B. M. de Waal. Oxford: Oxford University Press, 2002.

Kortlandt, Adriaan. "Chimpanzees in the Wild." *Scientific American* 206, no. 5 (1962): 128–38.

Kummer, Hans. "Le comportement social des singes." In *La recherche en éthologie: Les comportements animaux et humains,* edited by J.-P. Desportes and A. Vloebergh. Paris: Éd. du Seuil, coll. "Points Sciences / La Recherche," 1979.

———. *Social Organization of Hamadryas Baboons: A Field Study.* Basel: Karger, 1968.

Kuroda, Ryo. "On the Counting Ability of a Monkey (*Macacus cynomolgus*)." *Journal of Comparative Psychology* 12, no. 2 (1931): 171–80.

Kuroda, Suehisa. *Pygmy Chimpanzee: The Unknown Ape.* Tokyo: Chikuma Shobō, 1982.

Lamarck, Jean-Baptiste. *Zoological Philosophy: An Exposition with Regard to the Natural History of Animals.* Chicago: University of Chicago Press, 1984.

Lancaster, Jane B. *Primate Behavior and the Emergence of Human Culture.* New York: Holt, Rinehart & Winston, 1975.

Latour, Bruno. *Pandora's Hope: Essays on the Reality of Science Studies.* Cambridge, MA: Harvard University Press, 1999.

———. *Politics of Nature: How to Bring the Sciences into Democracy.* Cambridge, MA: Harvard University Press, 2004.

———. *We Have Never Been Modern.* Cambridge, MA: Harvard University Press, 1993.

Leakey, Louis S. B. *Adam's Ancestors: The Evolution of Man and His Culture.* New York: Harper & Row, 1960.

Le Gros Clark, Wilfrid Edward. *Early Forerunners of Man.* Baltimore: W. Wood, 1934.

Lenain, Thierry. *Monkey Painting.* London: Reaktion Books, 1997.

Lestel, Dominique. *Les animaux sont-ils intelligents?* Paris: Le Pommier, 2006.

———. *Paroles de singes: L'impossible dialogue homme-primate.* Paris: Éditions de la Découverte, 1995.

Ley, Ronald. *A Whisper of Espionage: Wolfgang Köhler and the Apes of Tenerife.* New York: Avery, 1990.

Linden, Eugene. *Apes, Men, and Language.* New York: Saturday Review, 1974.

———. "The Gambia: Signs in the Wilderness." *Atlantic,* March 1986, 32–37.

Linnaeus, Carl von. *Systema Naturae.* 2 vols. 10th ed. Stockholm: Impesis Laurentii Salvii, 1758.

Linnaeus, Carl von, and Christianus Emmanuel Hoppius. "Anthropomorpha." In *Dissertatio Academica.* Uppsala, 1760.

Lintz, Gertrude A. Davies. *Animals Are My Hobby.* New York: Robert M. McBride, 1942.

Lonsdorf, Elizabeth V., Stephen R. Ross, and Tetsuro Matsuzawa. *The Mind of the Chimpanzee.* Chicago: University of Chicago Press, 2010.

Lorenz, Konrad Z. *On Aggression.* New York: Harcourt, Brace & World, 1966.

Lykknes, Annette, Donald L. Optiz, and Brigitte Van Tiggelen, eds. *For Better or for Worse? Collaborative Couples in the Sciences.* Basel: Springer, 2012.

Mackenzie, Louisa, and Stephanie Posthumus, eds. *French Thinking About Animals: The Animal Turn.* East Lansing: Michigan State University Press, 2015.

MacKinnon, John Ramsay. *In Search of the Red Ape.* London: Collins, 1974.

Magnus, Albertus. *De Animalibus.* Venice: Joannes & Gregorius de Gregoriis, de Forlivio, 1495.

Marais, Eugène Nielen. *My Friends the Baboons.* London: Methuen, 1939.

———. *The Soul of the Ape.* New York: Atheneum, 1919.

Martinez-Contreras, Jorge. "L'émergence scientifique du gorille." *Revue de synthèse,* 4th ser., nos. 3–4 (1992).

Matsuzawa, Tetsuro. "Chimpanzee Ai and Her Son Ayumu: An Episode of Education by Master-Apprenticeship." In *The Cognitive Animal: Empirical and Theoretical Perspectives on Animal Cognition,* edited by Marc Bekoff, Colin Allen, and Gordon M. Burghardt. Cambridge: MIT Press, 2002.

———. *Primate Origins of Human Cognition and Behavior.* Tokyo: Springer, 2008.

Matsuzawa, Tetsuro, Tatyana Humle, and Sugiyama Yukimaru. *The Chimpanzees of Bossou and Nimba.* Tokyo: Springer, 2011.

Matsuzawa, Tetsuro, Shozo Kojima, and Masako Jitsumori. "Editorial: A Brief Note on the Historical Background of the Study of Cognition and Behavior in Nonhuman Primates by Japanese Researchers." *Japanese Psychological Research* 38, no. 3 (1996): 109–12.

Matsuzawa, Tetsuro, and William C. McGrew. "Kinji Imanishi and 60 Years of Japanese Primatology." *Current Biology* 18, no. 14 (2008).

Matthews, John. *A Voyage to the River Sierra-Leone on the Coast of Africa.* London: White & Sewell, 1788.

McDermott, William Coffman. "The Ape in Antiquity." *Johns Hopkins University Studies in Archaeology,* no. 27 (1938).

McGrew, William. *Chimpanzee Material Culture.* Cambridge: Cambridge University Press, 1992.

———. *The Cultured Chimpanzee.* Cambridge: Cambridge University Press, 2004.

McGrew, William C., Linda F. Marchant, and Toshisada Nishida, eds. *Great Ape Societies.* Cambridge: Cambridge University Press, 1996.

Menzel, Emil W. "Chimpanzee Spatial Memory Organization." *Science* 182 (1973): 943–45.

Merton, Robert K. "The Matthew Effect in Science." *Science* 159, no. 3810 (1968): 56–63.

Meunier, Victor. *Les singes domestiques.* Paris: Maurice Dreyfous, 1886.

Milne-Edwards, Henri. "The Young Gorilla of the Jardin des Plantes." *Nature,* June 5, 1884, 128–29.

Monboddo, James Burnett. *Of the Origin and Progress of Language.* 1773–1792. Reprint, New York: AMS Press, 1973.

Montgomery, Georgina M. "'Infinite Loneliness': The Life and Times of Miss Congo." *Endeavour* 33, no. 3 (2009): 101–5.

———. "Place, Practice and Primatology: Clarence Ray Carpenter, Primate Communication and the Development of Field Methodology, 1931–1945." *Journal of the History of Biology* 38 (2005): 495–533.

———. *Primates in the Real World: Escaping Primate Folklore and Creating Primate Science.* Charlottesville: University of Virginia Press, 2015.

Morbeck, Mary Ellen, Alison Galloway, and Adrienne Zihlman. *The Evolving Female: A Life-History Perspective.* Princeton: Princeton University Press, 1997.

Morgan, Conwy Lloyd. *An Introduction to Comparative Psychology.* 1st ed. London: Walter Scott, 1894.

———. *An Introduction to Comparative Psychology.* 2nd ed. London: Walter Scott, 1903.

———. "The Limits of Animal Intelligence." In *International Congress of Experimental Psychology,* 44–48. London: Williams & Norgate, 1892.

Morin, Alain. "Self-Recognition, Theory-of-Mind, and Self-Awareness: What Side Are You On?" *Laterality* 16, no. 3 (2011): 367–83.

Morris, Desmond. *The Biology of Art: A Study of the Picture-Making Behaviour of the Great Apes and Its Relationship to Human Art.* New York: Knopf, 1962.

Morris, Desmond, and Ramona Morris. *Men and Apes.* New York: McGraw-Hill, 1966.

Moussaief Masson, Jeffrey, and Susan McCarthy. *When Elephants Weep: The Emotional Lives of Animals.* New York: Delacorte, 1995.

Nagel, Thomas. "What Is It Like to Be a Bat?" *Philosophical Review* 83, no. 4 (1974): 435–50.

Nishida, Toshisada. *The Chimpanzee of the Mahale Mountains.* Tokyo: University of Tokyo Press, 1990.

Nissen, Henry Wieghorst. *A Field Study of the Chimpanzee: Observations of Chimpanzee Behavior and Environment in Western French Guinea. Comparative Psychology Monographs* 8, no. 1. Edited by Knight Dunlap. Baltimore: Johns Hopkins Press, 1931.

Noell, Anna Mae, and David K. Himber. *The History of Noell's Ark Gorilla Show.* Tarpon Springs, FL: Noell's Ark, 1979.

Ohnuki-Tierney, Emiko. "Representations of the Monkey." In Corbey and Theunissen, *Ape, Man, Apeman.*

Owen, Richard. "On the Osteology of the Chimpanzee and Orang Utan." *Transactions of the Zoological Society of London* 1 (1835): 343–79.

Parker, Sue Taylor, Robert W. Mitchell, and Lyn H. Miles. *The Mentalities of Gorillas and Orangutans: Comparative Perspectives.* Cambridge: Cambridge University Press, 1999.

Patterson, Francine, and Eugene Linden. *The Education of Koko: Can Animals Talk to Us?* New York: Henry Holt, 1988.

Pepys, Samuel. *The Diary of Samuel Pepys, Completely Transcribed by Minors Bright, from the Shortland Manuscript in the Pepysian Library, Magdalene College, Cambridge, with Lord Braybrooke's Notes.* [1660–69.] London: Henry B. Wheatley & G. Bell, 1893–99.

Perrot, Michelle. *Mon histoire des femmes.* Paris: Éd. du Seuil, coll. "Points Histoire," 2008.

Peterson, Dale. *Jane Goodall: The Woman Who Redefined Man.* Boston: Houghton Mifflin, 2006.

Pigafetta, Filippo. *A Report of the Kingdom of Congo, and the Surrounding Coun-*

tries; Drawn out of the Writings and Discourses of the Portuguese, Duarte Lopez. 1591. Reprint, New York: Negro Universities Press, 1969.

Pitman, Charles Robert S. "The Gorillas of the Kayonsa Region, Western Kigezi, S.W. Uganda." *Proceedings of the General Meeting for Scientific Business of the Zoological Society of London,* February 5, 1935, 477–94.

Plato. *Hippias Major.* Cambridge, MA: Harvard University Press, Loeb Classical Library, 1926.

Pliny the Elder. *Natural History.* Cambridge, MA: Harvard University Press, 2006.

Plowden, Gene. *Gargantua: Circus Star of the Century.* New York: Bonanza Books, 1972.

Polk, Milbry, and Mary Tiegreen. *Women of Discovery: A Celebration of Intrepid Women Who Explored the World.* New York: Clarkson Potter, 2001.

Pouillard, Violette. "Vie et mort des gorilles de l'Est (*Gorilla beringei*) en captivité (1923–2012)." *Revue de synthèse,* 6th ser., 136 (2015): 375–402.

Poussielgue, Achille. *Homme ou singe.* Paris: E. Dentu Libraire-Éditeur, 1861.

Povinelli, Daniel. *Folk Physics for Apes: The Chimpanzee's Theory of How the World Works.* Oxford: Oxford University Press, 2000.

Power, Margaret. *The Egalitarians, Human and Chimpanzee: An Anthropological View of Social Organization.* Cambridge: Cambridge University Press, 1991.

Pradhan, Gauri R., Maria A. van Noordwijk, and Carel van Schaik. "A Model for the Evolution of Developmental Arrest in Male Orangutans." *American Journal of Physical Anthropology* 149, no. 1 (2012): 18–25.

Premack, Ann James. *Why Chimps Can Read.* New York: Harper & Row, 1976.

Premack, David, and Ann James Premack. *The Mind of an Ape.* New York: Norton, 1984.

Premack, David, and Guy Woodruff. "Does the Chimpanzee Have a Theory of Mind?" In "Cognition and Consciousness in Nonhuman Species." Special issue, *Behavioral and Brain Sciences* 1, no. 4 (1978): 515–26.

Pruetz, Jill D., and Paco Bertolani. "Savanna Chimpanzees, *Pan troglodytes verus,* Hunt with Tools." *Current Biology,* March 6, 2007, 412–17.

Pycior, Helen M., Nancy G. Slack, and Pnina G. Abir-Am, eds. *Creative Couples in the Sciences.* New Brunswick, NJ: Rutgers University Press, 1996.

Radick, Gregory. *The Simian Tongue: The Long Debate About Animal Language.* Chicago: University of Chicago Press, 2007.

Raven, Henry C. *The Anatomy of the Gorilla.* New York: Columbia University Press, 1950.

———. "Further Adventures of Meshie." *Natural History* 33, no. 6 (1933): 607–17.

———. "Gorilla: The Greatest of All Apes." *Natural History* 31, no. 3 (1931): 231–42.

———. *Meshie, Child of a Chimpanzee.* 1932; American Museum of Natural History, 1985. DVD.

———. "Meshie: The Child of a Chimpanzee; A Creature of the African Jungle Emigrates to America." *Natural History* (American Museum of Natural History) 32, no. 2 (1932): 158–66.

Read, Carveth. *The Origin of Man and His Superstitions.* Cambridge: Cambridge University Press, 1920.

Rees, Amanda. *The Infanticide Controversy: Primatology and the Art of Field Science.* Chicago: University of Chicago Press, 2009.

Renck, Jean-Luc, and Véronique Servais. *L'éthologie: Histoire naturelle du comportement.* Paris: Éd. du Seuil, coll. "Points Sciences," 2002.

Reynolds, Vernon. *Budongo: An African Forest and Its Chimpanzees.* New York: Natural History Press, 1965.

———. *The Chimpanzees of the Budongo Forest: Ecology, Behaviour, and Conservation.* Oxford: Oxford University Press, 2005.

———. "On the Identity of the Ape Described by Tulp in 1641." *Folia Primatologica* 5 (1967): 80–87.

Rich, Jeremy. *Missing Links: The African and American Worlds of R. L. Garner, Primate Collector.* Athens: University of Georgia Press, 2012.

Richard, Alison F. *Behavioral Variation: Case Study of a Malagasy Lemur.* Lewisburg, PA: Bucknell University Press, 1978.

Rijksen, H. D. *A Field Study of Sumatran Orang Utans (*Pongo pygmaeus abelii, *Lesson 1872): Ecology, Behavior, and Conservation.* Wageningen, the Netherlands: Mededlingen Landbouwhogeschool, H. Veenman & B. V. Zonen, 1978.

Ripley, Suzanne. "The Leaping of Langurs: A Problem in the Study of Locomotor Adaptation." *American Journal of Physical Anthropology* 26, no. 2 (1967): 149–70.

Romanes, George John. *Animal Intelligence.* New York: D. Appleton, 1883.

———. "Conscience in Animals." *Popular Science Monthly,* May 1876.

———. *Mental Evolution in Animals.* New York: D. Appleton, 1884.

———. "On the Mental Faculties of the Bald Chimpanzee (*Anthropopithecus calvus*)." *Proceedings of the Zoological Society of London* (1889): 316–21.

Rossiianov, Kirill. "Beyond Species: Il'ya Ivanov and His Experiments on Cross-

Breeding Humans with Anthropoid Apes." *Sciences in Context* 15, no. 2 (2002): 277–316.

Rossiter, Margaret W. "The Matthew Matilda Effect in Science." *Social Studies of Science* 23, no. 2 (1993): 325–41.

Rothmann, Max. "Über die Errichtung einer Station zur psychologischen und hirnphysiologischen Erforschung der Menschenaffen." *Berliner klinische Wochenshrift* 49 (1912): 1981–85.

Rowell, Thelma. "The Concept of Social Dominance." *Behavioral Biology* 2, no. 131 (1974).

———. *Social Behavior of Monkeys.* Harmondsworth, UK: Penguin Books, 1972.

Royer, Clémence. "La domestication des singes." *Revue d'anthropologie,* 3rd ser., 10 (1887).

———. "Facultés mentales et instincts sociaux des singes." *La revue scientifique,* August 1886.

Ruch, Theodore C. *Bibliographia Primatologica: A Classified Bibliography of Primates Other Than Man, Part I.* Springfield, IL: CC Thomas, 1941.

Ruiz, Gabriel, and Natividad Sánchez. "Wolfgang Köhler's *The Mentality of Apes* and the Animal Psychology of His Time." *Spanish Journal of Psychology* 17, no. E69 (2014): 1–25.

Russell, Tuttle H. *Apes and Human Evolution.* Cambridge, MA: Harvard University Press, 2014.

Russon, Anne E. *Orangutans: Wizards of the Rain Forest.* Toronto: Key Porter Books, 1999.

Savage, Thomas S., and Jeffries Wyman. "Notice of the External Characters and Habits of *Troglodytes gorilla,* a New Species of Orang from the Gaboon River." *Proceedings of the Boston Society of Natural Society* 2 (1847): 45–247.

Savage-Rumbaugh, Sue, and Roger Lewin. *Kanzi: The Ape at the Brink of the Human Mind.* New York: John Wiley & Sons, 1994.

Schaller, Georges B. *The Mountain Gorilla: Ecology and Behavior.* Chicago: University of Chicago Press, 1963.

———. *The Year of the Gorilla.* Chicago: University of Chicago Press, 1964.

Schiebinger, Londa. *Nature's Body: Gender in the Making of Modern Science.* New Brunswick, NJ: Rutgers University Press, 1993.

Schwarz, Ernst. "Das Vorkommen des Schimpansen auf den linken Kongo-Ufer." *Revue de zoologie et de botanique africains* 16 (April 1929): 425–26.

Sebastiani, Silvia. "L'orang-outan, l'esclave et l'humain: Une querelle des corps en régime colonial." *L'atelier: Revue électronique du CRH* 11 (2013).

Seville, Isidore of. *Etymologiae.* Books 11 and 12. Cambridge: Cambridge University Press, 2006.

Seyfarth, Robert M., and Dorothy L. Cheney. "The Origin of Meaning in Animal Signals." In "Mechanisms & Function." Special issue, *Animal Behaviour* 30 (2016): 1–8.

Skinner, Burrhus F. "A Case History in Scientific Method." *American Psychologist* 11, no. 5 (1956): 221–33.

Slocum, Sally. "Woman the Gatherer: Male Bias in Anthropology." In *Toward an Anthropology of Women,* edited by Rayna R. Reiter, 36–50. New York: Monthly Review, 1975.

Smuts, Barbara B. *Sex and Friendship in Baboons.* New York: Aldine, 1985.

Smuts, Barbara B., Dorothy L. Cheney, Robert M. Seyfarth, Richard W. Wrangham, and Thomas T. Struhsaker, eds. *Primates Societies.* Chicago: University of Chicago Press, 1986.

Stoczkowski, Wictor. "Portrait de l'ancêtre en singe: L'hominisation sans évolutionnisme dans la pensée naturaliste du XVIIIe siècle." In Corbey and Theunissen, *Ape, Man, Apeman.*

Strickland, Debra Higgs. *Saracens, Demons, and Jews: Making Monsters in Medieval Art.* Princeton: Princeton University Press, 2003.

Strivay, Lucienne. *Enfants sauvages: Approches anthropologiques.* Paris: Gallimard, 2006.

Struhsaker, Thomas T. *Behavior of Vervet Monkeys (*Cercopithecus aethiops*).* Berkeley: University of California Press, 1967.

Strum, Shirley C. *Almost Human: A Journey into the World of Baboons.* New York: Random House, 1987.

Strum, Shirley C., and Linda M. Fedigan. *Primate Encounters: Models of Science, Gender, and Society.* Chicago: University of Chicago Press, 2000.

Strum, Shirley C., Donald G. Lindburg, and David Hamburg, eds. *The New Physical Anthropology: Science, Humanism, and Critical Reflection.* Upper Saddle River, NJ: Prentice Hall, 1999.

Sugiyama, Yukimaru. "Influence of Provisioning on Primate Behavior and Primate Studies." *Mammalia* 79, no. 3 (2015): 255–65.

Susman, Randall L., ed. *The Pygmy Chimpanzee: Evolutionary Biology and Behavior.* New York: Plenum, 1984.

Tattersall, Ian. *Becoming Human: Evolution and Human Uniqueness.* Boston: Houghton Mifflin Harcourt, 1998.

———. *Masters of the Planet: The Search for Our Human Origins.* New York: St. Martin's, 2012.

Teleki, Geza. *The Predatory Behavior of Wild Chimpanzees.* Lewisburg, PA: Bucknell University Press, 1973.

Temerlin, Maurice K. *Lucy: Growing Up Human, A Chimpanzee Daughter in a Psychotherapist's Family.* Palo Alto, CA: Science and Behavior Books, 1975.

Terrace, Herbert S. *Nim: A Chimpanzee Who Learned Sign Language.* New York: Knopf, 1979.

Teuber, Marianne L. "The Founding of the Primate Station, Tenerife, Canary Islands." *American Journal of Psychology* 107, no. 4 (1994): 551–81.

Thinès, Georges, and René Zayan. "F. J. J. Buytendijk's Contribution to Animal Psychology or Ethology?" *Acta Biotheoretica* 24, nos. 3–4 (1975): 86–99.

Thomas, Marion. "Are Animals Just Noisy Machines? Louis Boutan and the Co-invention of Animal and Child Psychology in the French Third Republic." *Journal of the History of Biology* 38, no. 3 (2005): 425–60.

———. "Between Biomedical and Psychological Experiments: The Unexpected Connections Between the Pasteur Institutes and the Study of Animal Mind in the Second Quarter of the Twentieth-Century France." *Studies in History and Philosophy of Biological and Biomedical Sciences* 55 (2016): 29–40.

Thompson, Jo A. Myers. "The Status of Bonobos in Their Southernmost Geographic Range." In *All Apes Great And Small,* vol. 1., *African Apes,* edited by B. M. F. Galdikas, N. Briggs, L. K. Sheeran, G. L. Shapiro, and J. Goodall, 75–81. New York: Springer, 2001.

Thorndike, Edward L. *The Mental Life of the Monkeys.* Psychological Review Monograph Supplements 15. New York: Macmillan, 1901.

Tinbergen, Nikolaas. "On Aims and Methods of Ethology." *Zeitschrift für Tierpsychologie* 20 (1963): 410–33.

Tinklepaugh, Otto L. "Multiple Delayed Reaction with Chimpanzees and Monkeys." *Journal of Comparative Psychology* 13 (1932): 207–43.

Todes, Daniel P. *Ivan Pavlov: A Russian Life in Science.* Oxford: Oxford University Press, 2014.

Tolman, Edward C. "The Determiners of Behavior at a Choice Point." *Psychological Review* 45 (1938): 1–41.

———. *Purposive Behavior in Animals and Men.* New York: Appleton-Century, 1932.

Tomasello, Michael, and J. Call. *Primate Cognition.* New York: Oxford University Press, 1997.

Tulpius [Nicholas Tulp]. *Observationum Medicarum.* Book 4. Rev. ed. Amsterdam: Danielem Elzevirium, 1672.

Tyson, Edward. *Orang-Outang, sive* Homo sylvestris; *or, The Anatomy of a Pygmie Compared with That of a Monkey, an Ape, and a Man. To Which Is Added a Philological Essay concerning the Pygmies, the Cynocephali, the Satyrs and Sphinges of the Ancients. Wherein it Will Appear That They Are All Either Apes or Monkeys, and Not Men, as Formerly Pretended.* London: Thomas Bennet & Daniel Browne (& Mr. Hunt), 1699.

Uexküll, Jakob von. *Foray into the Worlds of Animals and Humans: With a Theory of Meaning.* Minneapolis: University of Minnesota Press, 2010.

van Schaik, Carel P. *Among Orangutans: Red Apes and the Rise of Human Culture.* Cambridge, MA: Belknap Press, Harvard University Press, 2004.

van Schaik, Carel P., Marc Ancrenaz, Gwendolyn Borgen, Biruté Galdikas, Cheryl D. Knott, Ian Singleton, Akira Suzuki, Sri Suci Utami, and Michele Merrill. "Orangutan Cultures and the Evolution of Material Culture." *Science* 299, no. 5603 (2003): 102–5.

Vauclair, Jacques. *Animal Cognition: An Introduction to Modern Comparative Psychology.* Cambridge, MA: Harvard University Press, 1996.

Vercors. *You Shall Know Them.* Boston: Little, Brown, 1953.

Vosmaer, Arnoet A. *Description de l'espèce de singe aussi singulier que très rare, nommé orang-outang, de l'isle de Borneo. Apporté vivant dans la ménagerie de Son Altesse Sérénissime, Monseigneur le Prince d'Orange et de Nassau, stadhouder héréditaire, gouverneur, capitaine général et amiral des Provinces-Unies des Pais-Bas, etc., etc., etc.* Amsterdam: Pierre Meijer, 1778.

Wallace, Alfred Russel. *The Malay Archipelago: The Land of the Orang-utan and the Bird of Paradise.* London: Macmillan, 1890.

Washburn, Sherwood L. *The New Physical Anthropology.* New York: New York Academy of Sciences, 1951.

Washburn, Sherwood L., and Chet S. Lancaster. "The Evolution of Hunting." In *Man the Hunter,* edited by Richard B. Lee, Irven DeVore, and Jill Nash-Mitchell. New York: Aldine De Gruyter, 1968.

Watson, John Broadus. "Psychology as the Behaviorist Views It." *Psychological Review* 20 (1913): 158–77.

Whiten, Andrew. "Evolving a Theory of Mind: The Nature of Non-verbal Mentalism in Other Primates." In *Understanding Other Minds: Perspectives from Autism,* edited by S. Baron-Cohen, H. Tager-Flusberg, and D. J. Cohen, 367–96. New York: Oxford University Press, 1993.

Whiten, Andrew, Jane Goodall, William C. McGrew, Toshisada Nishida, Vernon Reynolds, Yukimaru Sugiyama, Caroline E. G. Tutin, Richard W. Wrangham, and Christophe Boesch. "Cultures in Chimpanzees." *Nature* 399 (1999): 682–85.

Wilson, Edward Osborne. *Sociobiology: The New Synthesis.* Cambridge, MA: Harvard University Press, 1975.

Windholz, George. "Köhler's Insight Revisited." *Teaching of Psychology* 12, no. 3 (1985): 165–67.

———. "Pavlov vs. Köhler: Pavlov's Little-Known Primate Research." *Pavlovian Journal of Biological Science* 19, no. 1 (1984): 23–31.

Witmer, Lightner. "A Monkey with a Mind." *Psychological Clinic* 3 (1909): 179–205.

Wrangham, Richard W. *Chimpanzee Culture.* Cambridge, MA: Harvard University Press, 1994.

Wrangham, Richard W., and Dale Peterson. *Demonic Males: Apes and the Origins of Human Violence.* Boston: Mariner Books, 1996.

Wurmb, Friedrich von. "Description of the Large Orang-Outang of Borneo." *Philosophical Magazine* 1 (1798): 225–31.

Wyhe, John van, and Peter C. Kjaergaard. "Going the Whole Orang: Darwin, Wallace and the Natural History of Orangutans." *Studies in History and Philosophy of Biological and Biomedical Sciences* 51 (2015): 53–63.

Wynne, Clive D. L. "Rosalía Abreu and the Apes of Havana." *International Journal of Primatology* 29, no. 2 (2008): 289–302.

Yerkes, Robert M. *Almost Human.* New York: Century, 1925.

———. *Chimpanzees: A Laboratory Colony.* New Haven: Yale University Press, 1943.

———. "Creating a Chimpanzee Community." *Yale Journal of Biology and Medicine* 36, no. 219 (1963).

———. *The Mental Life of Monkeys and Apes. Behavioral Monographs* 3 (1916).

———. *The Mind of a Gorilla: Parts I & II. Genetic Psychology Monographs* 2 (1927).

———. *The Mind of a Gorilla: Part III. Comparative Psychology Monographs* 5 (1928).

———. "Provision for the Study of Monkeys and Apes." *Science* 43, no. 1103 (1916): 231–34.

Yerkes, Robert M., and Blanche W. Learned. *Chimpanzee Intelligence and Its Vocal Expressions.* Baltimore: Williams & Wilkins, 1925.

Yerkes, Robert M., and Ada W. Yerkes. *The Great Apes: A Study of Anthropoid Life.* New Haven: Yale University Press, 1929.

Zihlman, Adrienne L., and Debra R. Bolter. "Body Composition in *Pan paniscus* Compared with *Homo sapiens* Has Implications for Changes During Human Evolution." *Proceedings of the National Academy of Sciences* 112, no. 24 (2015): 7466–71.

Zuckerman, Solly. *The Social Life of Monkeys and Apes.* New York: Harcourt, Brace, 1922.

Index